DISCRETE MATHEMATICS
AND
ITS APPLICATIONS

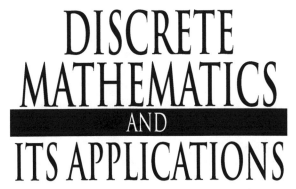

Series Editor

Kenneth H. Rosen, Ph.D.

T0144252

Continued Titles

DISCRETE MATHEMATICS AND ITS APPLICATIONS
Series Editor KENNETH H. ROSEN

AUTHENTICATION CODES AND COMBINATORIAL DESIGNS

DINGYI PEI

CRC Press
Taylor & Francis Group
Boca Raton London New York

CRC Press is an imprint of the
Taylor & Francis Group, an **informa** business
A CHAPMAN & HALL BOOK

CRC Press
Taylor & Francis Group
6000 Broken Sound Parkway NW, Suite 300
Boca Raton, FL 33487-2742

First issued in paperback 2019

© 2006 by Taylor & Francis Group, LLC
CRC Press is an imprint of Taylor & Francis Group, an Informa business

No claim to original U.S. Government works

ISBN-13: 978-1-58488-473-6 (hbk)
ISBN-13: 978-0-367-391249 (pbk)

Library of Congress Cataloging-in-Publication Data

Pei, Dingyi.
    Authentication codes and combinatorial designs / Dingyi Pei.
        p. cm.
    Includes bibliographical references and index.
    ISBN 1-58488-473-8 (9781584884736)
    1. Data encryption (Computer science) 2. Cryptography. 3. Combinatorial designs and configurations. I. Title.

QA76.9.A25P42 2005
005.8'2--dc22   2005026036

Visit the Taylor & Francis Web site at
http://www.taylorandfrancis.com

and the CRC Press Web site at
http://www.crcpress.com

Contents

Preface

This book contains original contributions of the author in the field of authentication schemes. Most of which are already published by international conferences and in journals, but they are presented with improvement in this book together with references and reviews of related work by other authors in the field.

Gus Simmons first introduced the concept of authentication schemes in the early 1980s as the cryptographic primitive for providing authentication in unconditionally secure systems. Gilbert, MacWilliams, and Sloane constructed one of the first authentication schemes. The original model of authentication assumed a sender wanted to send a message to a receiver over an insecure channel. The sender and the receiver were assumed to be trustworthy and shared a common key. An authentication scheme was used to allow the receiver to detect if a message was authentic or tampered with. Later, the trust assumption was reduced and an authentication scheme with four participants was used to provide protection when the sender and receiver were not trusted: the sender could deny a message that he had sent and the receiver could substitute a message that he had received. To protect against these attacks the sender and the receiver were given their individual keys. A trusted arbiter, with the knowledge of all the key information, was introduced into the model to arbitrate if a dispute between the sender and the receiver occurred.

Performance of authentication schemes in each of the above two models has been assessed using information-theoretic and combinatorial bounds on the probability of successful deception. There is a lower bound on the number of keys which is related to the probability of successful deception. It is proven that attainment of the lower bound on the number of keys implies attainment of the information-theoretic bounds on the probability of successful deceptions. Such schemes that achieve the lower bound on the number of keys will be called perfect. Perfect schemes have regular structures and can be characterized in terms of combinatorial designs.

It is proven that there is a connection between the perfect authentication schemes and the strong partially balanced designs ($SPBD$), which is a generalized concept of the well-known concept of t-designs. This fact establishes a bridge between the theory of authentication schemes and the theory of combinatorial designs. The t-design and the orthogonal array are two already known classes of SPBD. Using rational normal curves in the projective space over a finite field, a new class of SPBD constructed in this book.

By the simple, clean, and comprehensive language of mathematics, this book provides an excellent example to show how to reduce a problem from practice to a problem in mathematics. This result advances the study of authentication schemes, provides a new application of combinatorial designs in cryptography, and also raises a new research subject — how to construct SPBD in the theory of combinatorial designs.

This book is organized into the following nine chapters.

Chapter 1 introduces the concept of authentication schemes and the concept of combinatorial designs by some examples, while Chapter 2 describes the mathematical definition of the probability of successful deceptions.

Chapters 3 and 4 study the authentication schemes with three and four participants, respectively. The information-theoretic lower bounds on the probability of successful deceptions are obtained. Developing from this discussion, the lower bounds on the number of keys are derived. Then the combinatorial structure of perfect schemes, which achieve the lower bounds on the number of keys, is described. Special attention is focused on the perfect Cartesian schemes. The combinatorial bounds on the probability of successful deceptions are also discussed.

The focus of Chapter 5 is on the construction of the perfect authentication schemes based on the rational normal curves over finite fields. The composition of encoding rules for such schemes is explored for some special cases.

The subject of Chapters 6 through 8 is the presentation of some already known combinatorial designs which can be used to construct perfect (Cartesian) authentication schemes. They include *t*-designs, orthogonal arrays of index unity, and some designs constructed by finite geometry.

Chapter 9 studies the perfect authentication/secrecy schemes. Several definitions for perfect secrecy are introduced, and the properties of schemes with various types of perfect secrecy are studied. Some constructions of perfect secrecy schemes with or without authentication are also presented.

The Appendix contains a survey of constructions for authentication/secrecy schemes. Several constructions which are not mentioned in the previous chapters are also described.

The author would like to thank Dr. Bimal Roy for his help with proofreading, and Junwu Dong for his excellent typesetting.

Dingyi Pei

Chapter 1

Introduction

1.1 Authentication Problem

The importance of protecting the secrecy of sensitive messages has been realized by people since ancient times. Secret writing has been an established problem and this field was named cryptology, which was mainly concerned with military and diplomatic applications for a long time. By using the strong technique of information, the storage and transmission of information become cheap and simple in modern times. A huge amount of information is transformed in a way that almost anyone may access it. A lot of new problems related to cryptology appear. For example, an enemy might not only have the means to read transmitted messages, but could actually change them, or the enemy could produce and send a false message to the receiver and hope that this would initiate some action.

As an example, we imagine a thief who has access a bank's outgoing telephone line. When the thief visits the bank and deposits 100 dollars on his account, the bank sends a message to a central computer, telling the computer to add 100 dollars to the thief's account. By changing the content in the message, the thief can add a different amount, for example 1000 dollars, to his account. Another possibility would be to record the message transmitted by the bank and then send the same message several times, each time adding 100 dollars to the account. This example shows that it is necessary to have some mechanism to check that only messages sent by the bank will be accepted by the central computer. We say that the messages need to be authenticated.

When a receiver obtains a message, such as an electronic mail, he or she may be concerned about who is the real sender and whether the content of the message has been changed illegally by somebody in transmission. These are two major points that the authentication of messages takes care of. The authentication problem as well as the secrecy problem become the two important aspects of information security in modern times.

A secrecy scheme is a way of obtaining secrecy protection, while an authentication scheme is a way of obtaining authentication protection. The secrecy and the authentication are two different issues of information security. The secrecy can be essential in some situations and in others it is not. So an authentication scheme may be with or without secrecy. Messages transmitted between different units in a military organization often need both secrecy and authentication. Let us look at an example of authentication without secrecy. A company that owns a lot of parking meters has an employee who collects the money from the meters. The company wants to check that the employee does not put some of the collected money in his own pocket. This can be achieved if each parking meter authenticates the amount of money it contains, for example, it can produce a message on a piece of paper, which the employee has to show together with the collected money. Since the employee counts the money, he knows the amount of money. This is a situation where the message to be authenticated has already been known by the enemy. An authentication scheme without secrecy could be used in this case.

In many cases of authentication there are usually a transmitter, a receiver, and an enemy. The transmitter wants to send some messages to the receiver, while the enemy wants to deceive the receiver by changing the message sent by the transmitter or by producing a false message. In some cases the transmitter and the receiver may be the same person, for example, such as in the case of authentication of data files. A data file stored in memory may have sensitive content, such as the data for salaries to be paid, results of student exams, or similar data. We can authenticate the data files in order to discover any illegal changes in it.

If the transmitter and the receiver are not the same person, they may be trustworthy, which means that they do not try to cheat each other. But there are many situations when this is not the case. The two communicants may then try to cheat each other. Let us consider the case of the stock market. A customer sends electronic messages to his stockbroker, giving orders to sell or buy stocks. There is a possibility that disputes between the customer and the stockbroker will occur. If the customer has sent an order to buy some stocks that later decrease in value, he might deny having sent the order. On the other hand, the stockbroker might claim to have received an order to buy or sell stocks, when no such order was given. An authentication scheme for two nontrusting parties is needed. To solve such disputes, the communicants must prove their claim to a trusted third party, called the arbiter.

An authentication scheme is secure if the enemy's possibility of successful deception is very small. A scheme is said to be unconditionally secure if the security is independent of the computing power that an enemy may use, or to be computationally secure where the enemy is assumed to have limited computing power. A digital signature is an authentication scheme that is computationally secure, which depends on the assumption that certain difficult problems cannot be solved efficiently (such as factoring a large number, calculating a discrete logarithm). This book is entirely devoted to unconditionally secure authentication schemes (with or without secrecy).

1.2 Authentication Schemes

For simplicity we assume that the transmitter has only two possible messages to be sent: Attack and Withdraw. Denote these messages by A and W, respectively. The transmitter may send "0" to mean A and "1" to mean W, or vice versa. Thus he has two keys: key 1 and key 2. The resulting scheme is shown in Table 1.1.

Table 1.1

	0	1
key 1	A	W
key 2	W	A

The transmitter uses one of these two keys each time, and only the transmitter and the receiver know which key is used. When the enemy observes "0" or "1" sent by the transmitter, he does not know what it means since he does not know which key is being used. Hence this scheme can protect secrecy. Assume that the probability of choosing each key is $1/2$. The enemy tries to guess which message was sent and he will be successful with a probability $1/2$.

We should mention Kerkhoff's assumption before going further into discussion. This famous assumption is adopted as a standard assumption in all areas of cryptology. It states that the only parameter kept secret from the enemy is the active key. All other parameters, such as the structure of the system, probability distributions in the system, and so forth, are publicly known. The assumption is natural and necessary for the analysis of a system.

Before transmission, the transmitter and the receiver should choose one key. By Kerkhoff's assumption, the enemy knows the content of Table 1.1 but does not know which key is chosen at the moment. But the enemy still has a possibility to cheat. Before observing any messages sent from the transmitter, the enemy sends "0" and "1" to the receiver which will be accepted by the receiver as a genuine message, and the receiver will take some action. Thus the enemy causes an action, although he does not know it in advance. This kind of attack is called an impersonation attack.

There is another kind of attack which the enemy may launch. The enemy can change the message that he has observed from "0" to "1" or from "1" to "0", so the new message is still accepted by the receiver and will cause an opposite action to what the transmitter intended. This kind of attack is called a substitution attack.

The scheme of Table 1.1 does not have the authentication function. Let us call a information (A and W above), which the transmitter intends to send, as a source state, and a message ("0" and "1" above), which the transmitter really transmits, as an encoded message. By adding the number of encoded messages we can obtain a scheme with authentication. Take $\{A, W\}$ as the set of source states, as above, and $\{00, 01, 10, 11\}$ as the set of encoded messages. Consider the scheme given in Table 1.2.

Table 1.2

	00	01	10	11
key 1	A	–	W	–
key 2	A	–	–	W
key 3	–	A	W	–
key 4	–	A	–	W

There are four keys in the scheme. When key 1 is used, the encoded message "00" means A and "10" means W, while "01" and "11" are never taken. We call "00" and "10" as valid encoded messages of key 1 and "01" and "11" as invalid encoded messages of key 1. Similarly, this holds for all other keys. Each key has two valid encoded messages and two invalid encoded messages. The receiver only accepts the valid encoded messages of the key being used, and rejects any invalid encoded message. Each encoded message is valid for two keys among all four keys. Assume that the probability of choosing any key is the same $(1/4)$. When the enemy launches an impersonation attack, he picks up an encoded message, for example "00", and sends it to the receiver. The probability that "00" will be accepted by the receiver, i.e., the probability that "00" is valid of the key being used at the moment, is $1/2$. When the enemy launches a substitution attack, he tampers the encoded message sent by the transmitter and replaces it by a fraudulent one. Suppose the encoded message that he has observed is "00", thus he knows that the key 1 or key 2 has been chosen and "10" or "11" are two other possible valid encoded messages. Then he picks up one of them as his fraudulent encoded message and sends it to the receiver. His successful probability is also $1/2$. Thus, the scheme of Table 1.2 has the authentication function. We see from Table 1.2 that the encoded messages "00" and "01" are always used to transmit the source state A while "10" and "11" are always used to transmit the source state W. So anybody who observes the encoded message sent by the transmitter knows the corresponding source state. This is a scheme without secrecy.

The secrecy can be obtained if we change the scheme of Table 1.2 into the one given in Table 1.3. Examining the scheme given in Table 1.3 as has been done for Table 1.2 above,

Table 1.3

	00	01	10	11
key1	A	–	W	–
key2	W	–	–	A
key3	–	W	A	–
key4	–	A	–	W

we see that both the impersonation attack and the substitution attack succeed with a probability of $1/2$. This is a scheme with secrecy, since each encoded message can be used to transmit both A and W. The enemy can guess which source state was transmitted with a successful probability of $1/2$.

1.3 Combinatorial Designs

Combinatorial design theory concerns questions about whether it is possible to arrange elements of a finite set into subsets so that certain "balance" properties are satisfied. A subset is also called a block in the combinatorial design theory.

Each authentication scheme corresponds to a combinatorial design. Consider the scheme of Table 1.2. Let $\mathcal{M} = \{00, 01, 10, 11\}$ be the set of all possible encoded messages. Each key has a set of its valid encoded messages which is a subset of $\mathcal{M}$. Thus we have four such subsets of $\mathcal{M}$:

$$\mathcal{B} = \{(00, 10), (00, 11), (01, 10), (01, 11)\}.$$

We can see that each encoded message appears in two subsets of $\mathcal{B}$. The pair of sets $(\mathcal{M}, \mathcal{B})$ is called a combinatorial design. The scheme of Table 1.3 corresponds to the same combinatorial design $(\mathcal{M}, \mathcal{B})$ as the scheme of Table 1.2 does.

A mathematical description of authentication schemes will be given in Chapter 2. In the general model of authentication schemes, each key (it will be called an encoding rule there) has a set of valid encoded messages which is a subset of the set of all possible encoded messages denoted by $\mathcal{M}$. Let $\mathcal{E}$ denote the set of all keys and $\mathcal{M}(e)$ denote the set of valid encoded messages for the key $e \in \mathcal{E}$. Denote the collection of all such $\mathcal{M}(e)$ for all $e \in \mathcal{E}$ by $\mathcal{B}$. When we require that the authentication scheme possesses some desired properties, the corresponding pair $(\mathcal{M}, \mathcal{B})$ will have some balance properties and becomes a combinatorial design of a certain type. Conversely, if we have a combinatorial design of a certain type, then based on this design we can construct an authentication scheme with some desired properties (see Theorems 3.1 and 4.1).

Let us look at some examples of combinatorial designs.

Example 1.1
  Let

$$\mathcal{M} = \{1, 2, 3, 4, 5, 6, 7\}$$

$$B = \{123, 145, 167, 246, 257, 347, 356\}.$$

each element of $\mathcal{M}$ appears in three blocks, and each pair of elements of $\mathcal{M}$ appears in one block. □

Example 1.2

Let

$$\mathcal{M} = \{0, 1, 2, 3, 4, 5, 6, 7, 8, 9\}$$

$$\mathcal{B} = \{0123, 0145, 0246, 0378, 0579, 0689, 1278,$$

$$1369, 1479, 1568, 2359, 2489, 2567, 3458, 3467\}.$$

each element of $\mathcal{M}$ appears in six blocks, and each pair of elements of $\mathcal{M}$ appears in two blocks. ▯

We call the combinatorial designs of Examples 1.1 and 1.2 as 2-designs (also called balanced incomplete block designs). Each pair of elements of $\mathcal{M}$ appears in the same number (1 or 2 in the examples) of blocks. This concept can be generalized to so-called t-designs for any positive integer t, when any t-subset of $\mathcal{M}$ appears in exactly the same number of blocks.

Example 1.3

Let

$$\mathcal{M} = \{1, 2, 3, 4, 5, 6, 7, 8\}$$

$$\mathcal{B} = \{1368, 1357, 1458, 1467, 2468, 2457, 2358, 2367\}.$$

▯

Each element of $\mathcal{M}$ appears in four blocks. Each pair of elements, except 12,34,56,78, appears in two blocks. The pairs $12, 34, 56, 78$, do not appear in any block. Each triple of elements, which does not contain any pair in $\{12, 34, 56, 78\}$, appears in one block, otherwise it does not appear in any block.

We call this kind of combinatorial design as partially balanced t-designs ($t = 3$ in this example), which will play an important role in the construction of perfect authentication schemes.

Design theory makes use of tools from linear algebra, finite field, finite geometry, number theory, etc. The basic concepts of design theory are quite simple, but the mathematics needed to study designs are varied, rich, and ingenious.

Chapter 2

Authentication Schemes

Secrecy and authentication are two fundamental aspects of information security. Secrecy provides protection for sensitive messages against eavesdropping by an unauthorized person, while authentication provides protection for messages against impersonating and tampering by an active deceiver. Secrecy and authentication are two independent concepts. It is possible that only secrecy, or only authentication, or both are concerned in an information system.

2.1   Model with Three Participants (A-Codes)

We consider the authentication model that involves three participants: a transmitter, a receiver, and an opponent. The transmitter wants to communicate a sequence of source states to the receiver. In order to deceive the receiver, the opponent impersonates the transmitter to send a fraudulent message to the receiver, or to tamper with the message sent to the receiver. The transmitter and the receiver must act with the common purpose to deal with the spoofing attack from the opponent. They are assumed to trust each other in this model. If the transmitter and the receiver may also cheat each other, it is necessary to introduce the fourth participant – an arbiter. The model with three participants is usually called an A-code and that with an arbiter is called an A^2-code in the literature. We introduce the A-code in this section, and the A^2-code in the next section.

Let $\mathscr{S}$ denote the set of all source states which the transmitter may convey to the receiver. In order to protect against attacks from the opponent, source states are encoded using one encoding rule. Let $\mathscr{E}$ denote the set of all encoding rules, and $\mathscr{M}$ denote the set of all possible encoded messages. Usually, the number of encoded messages is much larger than that of source states. An encoding rule $e \in \mathscr{E}$ is a one-to-one mapping from $\mathscr{S}$ to $\mathscr{M}$. The range of the mapping $e(\mathscr{S})$ is called the set of valid messages of e, which is a subset of $\mathscr{M}$. Prior to transmission the transmitter and the receiver agree upon an encoding rule e, which is kept secret from the opponent. The transmitter uses e to encode source states. The encoded messages are transmitted through a public insecure channel. When a message is received, the receiver checks whether it lies in the range $e(\mathscr{S})$. If it does, then the message is accepted as authentic, otherwise it is rejected. The receiver recovers the source state from the received message by determining its (unique) preimage under the agreed encoding rule e. We assume that the opponent has a complete understanding of the system, including all encoding rules. The only thing he does not know is the particular encoding rule agreed upon by the transmitter and the receiver. The opponent can be successful in his spoofing attack if and only if his fraudulent message is valid for the encoding rule used. The set of valid messages usually is different from rule to rule. In order to decrease the possibility of successful deception from the opponent, the used encoding rule must be alternated frequently.

We assume that the opponent has the ability to impersonate the transmitter to send messages to the receiver, or to tamper with the messages sent by the transmitter. This is called a spoofing attack of order r. After observation of the first r $(r \geq 0)$ messages sent by the transmitter using the same encoding rule, the opponent places a fraudulent message, which is different from the observed r messages, into the channel, attempting to make the receiver accept it as authentic. It is an impersonation attack when r is zero. Let P_r denote the expected probability of successful deception for an optimum spoofing attack of order r. We are going to find an expression for P_r.

We think of source states, encoded messages, and encoding rules as random variables denoted by S, M, and E, respectively, i.e., there are probability distributions on the sets $\mathscr{S}, \mathscr{M}$, and $\mathscr{E}$, respectively. Let $p(S = s)$ denote the probability of the event that the variable S takes the value $s \in \mathscr{S}$. Similarly we have $p(E = e)$ and $p(\mathscr{M} = m)$ where $e \in \mathscr{E}$ and $m \in \mathscr{M}$. For simplicity, we abbreviate by omitting the names of random variables in a probability distribution when this causes no confusion. For instance, we write $p(s), p(e)$, and $p(m)$ for the above probabilities. Similarly, we write the conditional probability $p(e \mid m)$ instead of $p(E = e \mid M = m)$.

For any $m^r = (m_1, m_2, \cdots, m_r) \in \mathscr{M}^r$ and $e \in \mathscr{E}$, let $f_e(m^r) = \{f_e(m_1), \cdots, f_e(m_r)\}$ denote the elements $(s_1, \cdots, s_r) \in \mathscr{S}^r$, when $m_i \in e(\mathscr{S})$ for $1 \leq i \leq r$, such that $s_i = f_e(m_i)$ is the preimage of m_i under e.

We consider only impersonation $(r = 0)$ and plaintext substitution. The latter means that the opponent is considered to be successful only when, after observing a sequence of messages $m_1, \cdots, m_r$, he chooses a fraudulent message m' that is valid for the used encoding rule e, and $f_e(m') \neq f_e(m_i), 1 \leq i \leq r$. In this case, the receiver is informed of a source state which the transmitter does not intend to convey. Furthermore, if the receiver gets a particular message twice, he has no way to decide whether the message was sent twice by the transmitter or was repeated by an opponent. Hence, we assume that the transmitter never sends the same source state twice using the same encoding rule. Here we also assume that for any tuple $s^r = (s_1, s_2, \cdots, s_r)$ (ordered) its probability of occurrence is positive only if the r source states $s_1, s_2, \cdots, s_r$ are distinct. Let $\mathscr{S}^r$ denote the set of all r-tuples of distinct source states and S^r be the random variable of the first r source states, which the transmitter intends to convey to the receiver, taking its values in $\mathscr{S}^r$. Furthermore, we assume that the random variable S^r has a positive distribution, which means that for every element $(s_1, \cdots, s_r)$ of $\mathscr{S}^r$,

$$p(S^r = (s_1, \cdots, s_r)) > 0. \tag{2.1}$$

The reason of the assumption will be explained in the remark after Proposition 3.1.

We require that for each encoding rule e,

$$p(e) > 0. \tag{2.2}$$

This can be achieved by simply removing those encoding rules with zero probability of occurrence from the set $\mathscr{E}$.

Assume that the random variables E and S^r are independent, i.e.,

$$p(e, s^r) = p(e)p(s^r), \quad \text{for all } e \in \mathscr{E}, \ s^r \in \mathscr{S}^r.$$

Let r-tuple $m^r = (m_1, \cdots, m_r) \in \mathscr{M}^r$, where $m_i, 1 \leq i \leq r$ are distinct from each other. Define

$$\mathscr{E}(m^r) = \{e \in \mathscr{E} \mid m_i \in e(\mathscr{S}), 1 \leq i \leq r\}.$$

The set $\mathscr{E}(m^r)$ may be empty. We require that $\mathscr{E}(m)$ is not empty for each $m \in \mathscr{M}$, otherwise the message is never used; it can been dismissed from $\mathscr{M}$. We call r-tuple m^r allowable when $\mathscr{E}(m^r) \neq \emptyset$. For any positive integer r denote the set of all allowable r-tuples m^r by

$$\overline{\mathscr{M}^r} = \{m^r \in \mathscr{M}^r \mid \mathscr{E}(m^r) \neq \emptyset\}.$$

Let M^r denote the random variable of the first r messages sent by the transmitter. The probability distributions on $\mathscr{E}$ and on $\mathscr{S}^r$ determine the distribution on $\overline{\mathscr{M}^r}$. For any $e \in \mathscr{E}$ and $m^r \in \overline{\mathscr{M}^r}$, if $e \notin \mathscr{E}(m^r)$, then $p(e, m^r) = 0$; if $e \in \mathscr{E}(m^r)$, then

$$p(e, m^r) = p(e)p(f_e(m^r)). \tag{2.3}$$

Let $P(m \mid m^r)$ be the probability that the message m is valid given that m^r has been observed. Then

$$P(m \mid m^r) = \sum_{e \in \mathscr{E}(m^r * m)} p(e \mid m^r),$$

where $m^r * m$ denotes the message sequence $m_1, \cdots, m_r, m$. Given that m^r has been observed, the opponent's optimum strategy is to choose the message m' that maximizes $P(m \mid m^r)$. Thus, the unconditional probability of success in an optimum spoofing attack of order r is

$$P_r = \sum_{m^r \in \overline{\mathscr{M}^r}} p(m^r) \max_{m \in \mathscr{M}} P(m \mid m^r). \tag{2.4}$$

(There will be no summation when $r = 0$.)

Let p_S and p_E denote the probability distribution of S and E, respectively. We refer to the 5-tuple $(\mathscr{S}, \mathscr{M}, \mathscr{E}, p_S, p_E)$ as an authentication scheme, while the 3-tuple $(\mathscr{S}, \mathscr{M}, \mathscr{E})$ as an authentication code (A-code).

2.2   Model with Four Participants (A^2-Codes)

In the authentication model with three participants discussed in the previous section, the transmitter and the receiver are assumed trusted, they do not cheat each other. But it is not always the case: the transmitter could deny a message that he had sent, and the receiver could attribute a fraudulent message to the transmitter. A trusted third party, called the arbiter, is introduced.

Let $\mathscr{S}$ and $\mathscr{M}$ denote the set of source states and encoded messages, respectively, as above. We define encoding rules of the transmitter and decoding rules of the receiver as follows. An encoding rule is a one-to-one mapping from $\mathscr{S}$ to $\mathscr{M}$. Let $\mathscr{E}_T$ denote the set of all encoding rules. A decoding rule is a mapping from $\mathscr{M}$ onto $\mathscr{S} \cup \{\text{reject}\}$, each message corresponding to a source state or to "reject." In the former case, the message is called valid of the decoding rule and will be accepted by the receiver; in the latter case, the message will be rejected by the receiver. The set of all decoding rules is denoted by $\mathscr{E}_R$.

Suppose $f \in \mathscr{E}_R$ is a decoding rule of the receiver and $s \in \mathscr{S}$ is a source state, let $\mathscr{M}(f, s)$ denote the set of all valid messages of f corresponding to s. The sets $\mathscr{M}(f, s) \subset \mathscr{M}$ are disjointed for different source states s.

We say that an encoding rule $e \in \mathscr{E}_T$ is valid of the decoding rule $f \in \mathscr{E}_R$, or f is valid of e, if $e(s) \in \mathscr{M}(f, s)$ for any $s \in \mathscr{S}$.

Prior to transmission, the receiver selects a decoding rule $f \in \mathscr{E}_R$ and secretly gives it to the arbiter. The arbiter selects one message from each $\mathscr{M}(f, s)$ for all source states $s \in \mathscr{S}$, forming an encoding rule $e \in \mathscr{E}_T$ and secretly gives it to the transmitter to be used. In this case, the encoding rule e is valid of the decoding rule f. Receiving a message, the receiver checks whether it is a valid message of f, i.e., whether it is in some subset $\mathscr{M}(f, s)$; if it is then he accepts it as authentic and recovers the corresponding source state. When the transmitter and the receiver are disputing whether one message has been sent or not sent by the transmitter, the arbiter checks whether the message under dispute is valid of the encoding rule given to the transmitter. If it is valid, the arbiter thinks that the message is sent by the transmitter since only the transmitter knows the encoding rule. In the opposite case, the arbiter thinks that the message is not sent by the transmitter. Since the receiver does not know how the arbiter constructs the encoding rule e, it is not easy to choose a fraudulent message which is valid of e, that is, it is not easy to attribute a fraudulent message to the transmitter. Similarly, since the transmitter does not know the decoding rule f selected by the receiver, it is hard for the transmitter to choose a fraudulent message which is valid of f but is not valid of e, that is, it is hard for the transmitter to deny a message that he has sent. As for the opponent, he may be successful in a deception only if he can find a valid message of f.

The following three types of spoofing attacks are considered.

The spoofing attack O_r by the opponent: after observing a sequence of r distinct messages $m_1, m_2, \cdots, m_r$ sent by the transmitter under the same encoding rule, the opponent sends a message $m, m \neq m_i, 1 \leq i \leq r$ to the receiver and succeeds if the receiver accepts the message as authentic and the message is used to transmit a different source state from those transmitted by $m_i, 1 \leq i \leq r$.

The spoofing attack R_r by the receiver: after receiving a sequence of r distinct messages $m_1, m_2, \cdots, m_r$, the receiver claims to have received a different message m and succeeds if the message m is valid of the encoding rule used by the transmitter.

The spoofing attack T by the transmitter: the transmitter sends a message to the receiver and then denies having sent it. The transmitter succeeds if this message is accepted by the receiver as authentic and it is not valid of the encoding rule used by the transmitter.

Let P_{O_r}, P_{R_r}, and P_T denote the expected success probability for the optimal spoofing attack of the three kinds defined above, respectively. We are going to find their expressions.

Assume Equation (2.1) still holds. For a given decoding rule $f \in \mathscr{E}_R$, define

$$\mathscr{M}(f) = \bigcup_{s \in \mathscr{S}} \mathscr{M}(f, s)$$

and

$$\mathscr{E}_T(f) = \{e \in \mathscr{E}_T \mid e \text{ is valid of } f\}.$$

$\mathscr{M}(f)$ is the set of all valid messages for f. $\mathscr{E}_T(f)$ is the set of all encoding rules which are valid of f. We assume that for any given encoding rule $e \in \mathscr{E}_T$, there exists at least one decoding rule f such that $e \in \mathscr{E}_T(f)$.

For a given encoding rule $e \in \mathscr{E}_T$, let

$$\mathscr{M}(e) = \{e(s) \mid s \in \mathscr{S}\} \subset \mathscr{M}$$

and

$$\mathscr{E}_R(e) = \{f \in \mathscr{E}_R \mid e \text{ is valid of } f\}.$$

For a given $m^r = (m_1, \cdots, m_r) \in \mathscr{M}^r$, define the set

$$\mathscr{E}_R(m^r) = \{f \in \mathscr{E}_R \mid m_i \in \mathscr{M}(f), f(m_i) \neq f(m_j), 1 \leq i < j \leq r\}.$$

$\mathscr{E}_R(m^r)$ is the set of all decoding rules of which m^r is valid. For any given $f \in \mathscr{E}_R$, define

$$\mathscr{E}_T(f, m^r) = \{e | e \in \mathscr{E}_T(f), m_i \in \mathscr{M}(e), 1 \le i \le r\}.$$

This is the set of all encoding rules of which f and m^r are valid.

Similar to the previous section, let S^r and M^r denote the random variables of the first r source states and the first r encoded messages, respectively, and let E_R and E_T denote the random variables of decoding rules and encoding rules, respectively. We also assume that $p(E_R = f) > 0$ for any $f \in \mathscr{E}_R$ and $p(E_T = e \,|\, E_R = f) > 0$ for any $e \in \mathscr{E}_T(f)$. It follows immediately that $p(E_T = e) > 0$ for any $e \in \mathscr{E}_T$ and $p(E_R = f \,|\, E_T = e) > 0$ for any $f \in \mathscr{E}_R(e)$.

For any message $m \in \mathscr{M}$, we assume that there exists at least one decoding rule f such that $m \in \mathscr{M}(f)$, otherwise m is never used and can be dismissed from $\mathscr{M}$. Similarly, for any message $m \in \mathscr{M}(f)$ there exists at least one encoding rule $e \in \mathscr{E}_T(f)$ such that $m \in \mathscr{M}(e)$, otherwise m can be dismissed from $\mathscr{M}(f)$.

Let $P(m \,|\, m^r)$ denote the probability of the event that the message m is acceptable by the receiver given that the first r messages $m^r = (m_1, m_2, \cdots, m_r)$ have been observed, where $m_1, \cdots, m_r, m$ represent different source states. We have

$$P(m \,|\, m^r) = \sum_{f \in \mathscr{E}_R(m^r * m)} p(f \,|\, m^r).$$

Similar to Equation (2.4) we define

$$P_{O_r} = \sum_{m^r \in \mathscr{M}^r} p(m^r) \max_{m \in \mathscr{M}} P(m \,|\, m^r). \tag{2.5}$$

Let $P(m \,|\, f, m^r)$ denote the probability of the event that the message m could be valid of the encoding rule used by the transmitter given the decoding rule f and the first r messages $m^r = (m_1, \cdots, m_r)$. We have

$$P(m \,|\, f, m^r) = \sum_{e \in \mathscr{E}_T(f, m^r * m)} p(e \,|\, f),$$

and define

$$P_{R_r} = \sum_{f \in \mathscr{E}_R} p(f) \sum_{m^r \in \mathscr{M}^r} p(m^r \,|\, f) \max_{m \in \mathscr{M}} P(m \,|\, f, m^r)). \tag{2.6}$$

For a given $f \in \mathscr{E}_R$ and $e \in \mathscr{E}_T(f)$ define

$$\mathscr{M}'(e) = \mathscr{M} \setminus \mathscr{M}(e)$$

and

$$\mathscr{M}'_f(e) = \mathscr{M}(f) \setminus \mathscr{M}(e) \subset \mathscr{M}'(e).$$

Let $P(m' \mid e)$ denote the probability of the event that the message $m' \in \mathscr{M}'(e)$ is acceptable by the receiver given the encoding rule e. We have

$$P(m' \mid e) = \sum_{f \in \mathscr{E}_R(e,m')} p(f \mid e),$$

where

$$\mathscr{E}_R(e, m') = \{f \mid f \in \mathscr{E}_R(e), m' \in \mathscr{M}'_f(e)\}.$$

We define

$$P_T = \sum_{e \in \mathscr{E}_T} p(e) \max_{m' \in \mathscr{M}'(e)} P(m' \mid e). \tag{2.7}$$

Let p_S, p_{E_R}, and p_{E_T} denote the probability distribution of S, E_R, and E_T, respectively. We refer to the 7-tuple $(\mathscr{S}, \mathscr{M}, \mathscr{E}_R, \mathscr{E}_T, p_S, p_{E_R}, p_{E_T})$ as an authentication scheme with arbitration, while the 4-tuple $(\mathscr{S}, \mathscr{M}, \mathscr{E}_R, \mathscr{E}_T)$ as an authentication code (A^2-code) with arbitration.

2.3 Comments

Gilbert, MacWilliams, and Sloane published their paper *"Codes Which Detect Deception"* [12] in 1974, which is the landmark paper in the authentication theory. At the same time Simmons was independently working on the same subject and established the authentication model with three and four participants [41, 42].

The definitions of success probabilities for spoofing attacks may have a slight difference in some related literatures (it is due to the choice between the maximum value and the average value), but the difference is not essential. The definitions in this chapter follow from those in References [28] and [31].

Chapter 3

Authentication Schemes with Three Participants

We study authentication schemes with three participants in this chapter. When we say authentication schemes, we always mean the schemes with three participants in this chapter. The successful probability P_r of optimal spoofing attack of order r was introduced in Section 2.1. The information-theoretic lower bound of P_r and the necessary and sufficient conditions for achieving this bound will be given in Section 3.2 of this chapter. In order to discuss the information-theoretic bound, the important concept of information theory, entropy, is introduced in Section 3.1; some often used properties of entropy are proven. One important aim in constructing authentication schemes is to make P_r as small as possible. This aim is achieved at the expense of the use of a very large number of encoding rules. We also hope that the number of encoding rules can be as small as possible when we require P_r as small as possible. When $P_r, 0 \leq r < t$ achieve their information-theoretic bounds, the schemes are called t-fold key-entropy minimal, and when the number of encoding rules achieves its lower bound, the schemes are called (t-fold) perfect. It is proved in Section 3.3 that each perfect authentication scheme is always a key-entropy minimal and it corresponds to a strong partially balanced design (SPBD will be defined in Section 3.3), and vice versa, each SPBD can be used to construct a perfect authentication scheme. Thus, to construct perfect authentication schemes reduces to finding SPBD. We will construct a new family of SPBD, in order to construct a new family of perfect authentication schemes in Chapter 5.

3.1 Entropy

Suppose that all possible values (or states) of the variable X are $x_1, \cdots, x_n$, the probability of X taking x_i is denoted by $p(x_i) > 0$, hence

$$\sum_{i=1}^{n} p(x_i) = 1. \tag{3.1}$$

We call X as a random variable. If $n = 1$, then $p(x_1) = 1$ and X always takes the value x_1. Therefore X is totally determined. If the probabilities of X taking $x_i, 1 \le i \le n$, are all equal, i.e., $p(x_1) = p(x_2) = \cdots = p(x_n) = 1/n$, X is most undetermined. For general random variables, its indeterminacy is between these two cases. We have a quantity to measure the indeterminacy; it is called entropy. The entropy of a random variable X is defined as

$$H(X) = -\sum_{i=1}^{n} p(x_i) \log p(x_i).$$

Here the base of the logarithm function is 2.

LEMMA 3.1
We have $0 \le H(X) \le \log n$. If X is totally determined, then $H(X) = 0$; if $p(x_1) = \cdots = p(x_n) = 1/n$, then $H(X)$ achieves its maximum $\log n$.

PROOF   It is only necessary to prove the inequality $0 \le H(X) \le \log n$; the latter conclusion is trivial. We write p_i for $p(x_i)$. Let

$$F = -\sum_{i=1}^{n} p_i \log p_i - \lambda \sum_{i=1}^{n} p_i.$$

If

$$\frac{\partial F}{\partial p_i} = -\log(e p_i) - \lambda = 0, \quad i = 1, 2, \cdots, n$$

(here e is the base of the natural logarithm function), then $p_1, p_2, \cdots, p_n$ are all equal with the value $1/n$. Hence $H(X)$ achieves its maximum value $\log n$ in this case. The conclusion $H(X) \ge 0$ is obvious.   ☐

Based on the above lemma we can think that the entropy $H(X)$ measures the indeterminacy of X. If each value of X is represented by r bits and X takes any sequence $(a_1, a_2, \cdots, a_r)$ of r bits with equal probability, then $H(X) = \log 2^r = r$. We see that a bit can be taken as the unit of entropy. In the above example, the probability of X taking $(a_1, a_2, \cdots, a_r)$ is $1/2^r$, the logarithm $-\log(1/2^r) = r$ can be explained as the amount of information provided when the event takes place. When an event with probability p takes place, the amount of information it provides is $\log p^{-1}$. The entropy $H(X)$ can be explained as the expected amount of information provided when X takes a value.

Let X and Y be two random variables. Let $p(x, y)$ be the joint probability of $X = x$ and $Y = y$ simultaneously, and $p(y|x)$ be the conditional probability of $Y = y$ given $X = x$. We have

$$p(x, y) = p(x)p(y|x) = p(y)p(x|y). \tag{3.2}$$

If $p(x|y) = p(x)$ for any x and y, then $p(x,y) = p(x)p(y)$. We say that X and Y are independent in this case. Define the joint entropy of X and Y by

$$H(X,Y) = -\sum_{x,y} p(x,y) \log p(x,y),$$

where the summation runs through all pairs (x,y) with $p(x,y) > 0$ (in the following all summations have this restriction, so we will not mention it again). Define conditional entropy by

$$H(X\,|\,Y) = -\sum_{x,y} p(x,y) \log p(x\,|\,y),$$

and

$$H(Y\,|\,X) = -\sum_{x,y} p(x,y) \log p(y\,|\,x).$$

$H(X\,|\,Y)$ denotes the indeterminacy of X when Y is given. If X and Y are independent, then

$$H(X\,|\,Y) = -\sum_{x,y} p(x,y) \log p(x)$$

$$= -\sum_{x} \log p(x) \sum_{y} p(x,y)$$

$$= -\sum_{x} p(x) \log p(x) = H(X);$$

similarly, we also have $H(Y\,|\,X) = H(Y)$.

LEMMA 3.2

$$H(X,Y) = H(X) + H(Y\,|\,X) = H(Y) + H(X\,|\,Y).$$

PROOF It is only necessary to prove the first equality since $H(X,Y) = H(Y,X)$. We have

$$H(X) + H(Y\,|\,X) = -\sum_{x} p(x) \log p(x) - \sum_{x,y} p(x,y) \log p(y\,|\,x)$$

$$= -\sum_{x,y} p(x,y) \log \left(p(x)p(y\,|\,x)\right)$$

$$= -\sum_{x,y} p(x,y) \log p(x,y) = H(X,Y).$$

Here, Equation (3.2) is used. ∎

If X and Y are independent, it follows from Lemma 3.2 that $H(X,Y) = H(X) + H(Y)$.

LEMMA 3.3 *(Jensen inequality)*
Suppose $x_i > 0, p_i > 0, 1 \leq i \leq n$, and $p_1 + p_2 + \cdots + p_n = 1$. Then

$$\sum_{i=1}^{n} p_i \log x_i \leq \log \left(\sum_{i=1}^{n} p_i x_i\right).$$

The equality holds if and only if $x_i, 1 \leq i \leq n$ are all equal.

PROOF   It is equivalent to proving that

$$x_1^{p_1} x_2^{p_2} \cdots x_n^{p_n} \leq p_1 x_1 + p_2 x_2 + \cdots + p_n x_n.$$

The left side is the geometric mean value and the right side is the arithmetic mean value. This is a well-known inequality.   ⬚

LEMMA 3.4

$$H(X|Y) \leq H(X).$$

PROOF   We have

$$H(X|Y) - H(X) = -\sum_{x,y} p(x,y) \log p(x|y) + \sum_{x,y} p(x,y) \log p(x)$$

$$= \sum_{x,y} p(x,y) \log \frac{p(x)p(y)}{p(x,y)}.$$

It follows from Lemma 3.3 that

$$H(X \mid Y) - H(X) \leq \log \left(\sum_{x,y} p(x,y) \frac{p(x)p(y)}{p(x,y)}\right) = 0.$$

⬚

Lemma 3.4 means that the indeterminacy of X could decrease and could not increase when Y is given. One may find some information about X from the given Y. The decrease of indeterminacy $H(X) - H(X|Y)$ is the lost amount of information.

3.2   Information-Theoretic Bound

PROPOSITION 3.1

The inequality

$$P_r \geq 2^{H(E|M^{r+1}) - H(E|M^r)} \tag{3.3}$$

holds for any integer $r \geq 0$. The equality holds if and only if for any $m^r * m \in \mathcal{M}^{r+1}$ with $\mathcal{E}(m^r * m) \neq \emptyset$, the ratio

$$\frac{p(e|m^r)}{p(e|m^r * m)} \tag{3.4}$$

is independent of m^r, m, and $e \in \mathcal{E}(m^r * m)$. When this equality holds, the probability P_r equals the ratio in Equation (3.4).

REMARK 3.1   If $e \in \mathcal{E}(m^r * m)$, then by the assumptions in Equations (2.1), (2.2), and (2.3),

$$p(m^r)p(e|m^r) = p(e, m^r) = p(e)p(f_e(m^r)) > 0,$$

the numerator and the denominator of the fraction in Equation (3.4) are nonzero. This is the reason why we assume S^r has a positive distribution.   ▯

PROOF   Let M_{r+1} denote the random variable of the $(r+1)$-th message. For a given $m^r = (m_1, \cdots, m_r) \in \mathcal{M}^r$, let

$$\text{supp}(M_{r+1}, E|m^r) = \{(m, e)| e \in \mathcal{E}(m^r * m), m \neq m_i, 1 \leq i \leq r\}$$

denote the support of the conditional probability distribution of the random variable pair (M_{r+1}, E) conditional on $M^r = m^r$. Then underbounding a maximum by an average gives

$$\max_{m \in \mathcal{M}} P(m|m^r) \geq \sum_{m \in \mathcal{M}} p(M_{r+1} = m|m^r)P(m|m^r) \tag{3.5}$$

$$= \sum_{(m,e) \in \text{supp}(M_{r+1}, E|m^r)} p(M_{r+1} = m|m^r)p(e|m^r)$$

$$= \tilde{E}\left(\frac{p(M_{r+1} = m|m^r)p(e|m^r)}{p(M_{r+1} = m, e|m^r)}\right)$$

where $\widetilde{E}$ is the expectation of $\mathrm{supp}(M_{r+1}, E|m^r)$. By using the Jensen inequality, we obtain

$$
\begin{aligned}
\log \max_{m \in \mathcal{M}} P(m|m^r) &\geq \widetilde{E}\left(\log \frac{p(M_{r+1} = m|m^r)p(e|m^r)}{p(M_{r+1} = m, e|m^r)} \right) \qquad (3.6)\\
&= H(M_{r+1}, E|M^r = m^r) - H(M_{r+1}|M^r = m^r)\\
&\quad - H(E|M^r = m^r),
\end{aligned}
$$

where

$$
H(M_{r+1}, E|M^r = m^r) = - \sum_{(m,e)\in \mathrm{supp}(M_{r+1},E|m^r)} p(M_{r+1} = m, e|m^r) \log p(M_{r+1} = m, e|m^r),
$$

$$
H(M_{r+1}|M^r = m^r) = - \sum_{m : p(M_{r+1}=m|m^r)>0} p(M_{r+1} = m|m^r) \log p(M_{r+1} = m|m^r),
$$

$$
H(E|M^r = m^r) = - \sum_{e \in \mathscr{E}(m^r)} p(e|m^r) \log p(e|m^r).
$$

Finally, we make another use of the Jensen inequality to obtain

$$
\begin{aligned}
\log P_r &= \log \sum_{m^r \in \mathcal{M}^r} p(m^r) \max_{m \in \mathcal{M}} P(m|m^r)\\
&\geq \sum_{m^r \in \mathcal{M}^r} p(m^r) \log \max_{m \in \mathcal{M}} P(m|m^r) \qquad (3.7)\\
&\geq H(M_{r+1}, E|M^r) - H(M_{r+1}|M^r) - H(E|M^r)\\
&= H(E|M^{r+1}) - H(E|M^r).
\end{aligned}
$$

Lemma 3.2 is used in the last equality.

From the above derivation, we see that equality holds in this bound if and only if the following two conditions are satisfied:

(i) in order that equalities in Equations (3.5) and (3.7) hold, $P(m|m^r)$ is independent of those m and m^r with $p(M_{r+1} = m \,|\, m^r) > 0$;

(ii) in order that equalities in Equations (3.6) and (3.7) hold, for any $m^r *
m \in \mathcal{M}^{r+1}$ with $\mathscr{E}(m^r * m) \neq \emptyset$, the ratio

$$
\frac{p(M_{r+1} = m|m^r)p(e|m^r)}{p(M_{r+1} = m, e \,|\, m^r)} = \frac{p(e|m^r)}{p(e|m^r * m)}
$$

is independent of m, m^r, and $e \in \mathscr{E}(m^r * m)$.

Condition (i) can be deduced from condition (ii) since, if $p(M_{r+1} = m|m^r)$

> 0, then $\mathscr{E}(m^r * m) \neq \emptyset$ and

$$P(m|m^r) = \sum_{e \in \mathscr{E}(m' * m)} p(e|m^r)$$

$$= \frac{p(e|m^r)}{p(e|m^r * m)} \sum_{e \in \mathscr{E}(m' * m)} p(e|m^r * m)$$

$$= \frac{p(e|m^r)}{p(e|m^r * m)}$$

This completes the proof of the Proposition 3.1.　　　　　　　　　　　　⬜

The lower bound of P_r in Proposition 3.1 is called the information-theoretic bound.

3.3　Perfect Authentication Schemes

By Proposition 3.1 we know that

$$P_0 P_1 \cdots P_{r-1} \geq 2^{H(E|M')-H(E)} \geq 2^{-H(E)}, \tag{3.8}$$

hence by Lemma 3.1 we have

$$|\mathscr{E}| \geq 2^{H(E)} \geq (P_0 P_1 \cdots P_{r-1})^{-1}. \tag{3.9}$$

DEFINITION 3.1　*An authentication scheme is called t-fold key-entropy minimal if its $P_r, 0 \leq r \leq t-1$ achieve their information-theoretic bounds, i.e.,*

$$P_r = 2^{H(E|M'^{+1})-H(E|M')}, \quad 0 \leq r \leq t-1$$

DEFINITION 3.2　*An authentication scheme is called t-fold perfect if the number of encoding rules $|\mathscr{E}|$ achieves its lower bound $(P_0 P_1 \cdots P_{t-1})^{-1}$.*

COROLLARY 3.1
An authentication scheme $\mathscr{A} = (\mathscr{S}, \mathscr{M}, \mathscr{E}, p_S, p_E)$ is t-fold perfect if and only if $\mathscr{A}$ is t-fold key-entropy minimal, p_E is uniform, and $H(E|M^t) = 0$.

PROOF　Suppose that $\mathscr{A}$ is t-fold perfect. Then $\mathscr{A}$ is t-fold key-entropy minimal, $|\mathscr{E}| = 2^{H(E)}$, and $H(E|M^t) = 0$ by Equations (3.8) and (3.9). It follows from Lemma 3.1 that p_E is uniform.

Conversely, assume that $\mathscr{A}$ is t-fold key-entropy minimal, p_E is uniform, and $H(E|M^t) = 0$. Then

$$P_r = 2^{H(E|m^{r+1}) - H(E|M^r)}, \quad 0 \le r \le t - 1$$

and

$$P_0 P_1 \cdots P_{t-1} = 2^{H(E|M^t) - H(E)} = 2^{-H(E)}.$$

Since p_E is uniform,

$$|\mathscr{E}| = 2^{H(E)} = (P_0 P_1 \cdots P_{t-1})^{-1}.$$

The result follows.   ☐

We study the characterization of the perfect authentication schemes and find its construction method in this section.

For any positive integer r define

$$\overline{\mathscr{M}^r} = \{m^r \in \mathscr{M}^r \mid \mathscr{E}(m^r) \ne \emptyset\}.$$

We assume that $\overline{\mathscr{M}^1} = \mathscr{M}$, otherwise some messages are never used and can be dismissed. Note that the probability distribution of the random variable M^r is really defined on the set $\overline{\mathscr{M}^r}$. For any $m^r \in \overline{\mathscr{M}^r}$ define

$$\mathscr{M}(m^r) = \{m \in \mathscr{M} \mid \mathscr{E}(m^r * m) \ne \emptyset\}.$$

COROLLARY 3.2
Suppose that $|\mathscr{S}| = k$, $|\mathscr{M}| = v$, the positive integer $t \le k$, and $(\mathscr{S}, \mathscr{M}, \mathscr{E}, p_S, p_E)$ is t-fold key-entropy minimal.

(i) *For any $m^r \in \overline{\mathscr{M}^r}$, $1 \le r \le t$ the probability $p(f_e[m^r])$ is independent of $e \in \mathscr{E}(m^r)$;*

(ii) *For any $m^r * m \in \overline{\mathscr{M}^{r+1}}, 0 \le r \le t - 1$, we have*

$$P_0 = \sum_{e \in \mathscr{E}(m)} p(e), \qquad P_r = \frac{\sum_{e \in \mathscr{E}(m^r * m)} p(e)}{\sum_{e \in \mathscr{E}(m^r)} p(e)},$$

therefore

$$\sum_{e \in \mathscr{E}(m^r)} p(e) = P_0 P_1 \cdots P_{r-1},$$

for any $m^r \in \overline{\mathscr{M}^r}$, $1 \le r \le t$;

(iii) *$P_0 = k/v$ and $|\mathscr{M}(m^r)| = (k - r)P_r^{-1}$ for any $m^r \in \overline{\mathscr{M}^r}$;*

(iv) *$|\overline{\mathscr{M}^r}| = C_k^r (P_0 P_1 \cdots P_{r-1})^{-1}, 2 \le r \le t$. (It is trivial that $|\overline{\mathscr{M}^1}| = v$.)*

PROOF If $e \in \mathcal{E}(m^r * m)$, we have, from Proposition 3.1, that

$$
\begin{aligned}
P_r &= \frac{p(e \mid m^r)}{p(e \mid m^r * m)} \\
&= \frac{p(e, m^r)p(m^r * m)}{p(e, m^r * m)p(m^r)} \\
&= \frac{p(f_e(m^r)) \sum_{e' \in \mathcal{E}(m' * m)} p(e')p(f_{e'}(m^r * m))}{p(f_e(m^r * m)) \sum_{e' \in \mathcal{E}(m^r)} p(e')p(f_{e'}(m^r))}.
\end{aligned}
\tag{3.10}
$$

Taking $r = 0$ in Equation (3.10) we obtain

$$
P_0 = \frac{\sum_{e' \in \mathcal{E}(m)} p(e')p(f_{e'}(m))}{p(f_e(m))}.
$$

It follows that $p(f_e[m])$ does not depend on $e \in \mathcal{E}(m)$, (i) holds for $r = 1$ and (ii) holds for $r = 0$. Using Equation (3.10), (i) and (ii) can be proved by induction on r.

We have

$$
vP_0 = \sum_{m \in \mathcal{M}} \sum_{e \in \mathcal{E}(m)} p(e) = k \sum_{e \in \mathcal{E}} p(e) = k,
$$

hence, the first equality of (iii) holds. For any given $m^r = (m_1, \cdots, m_r) \in \overline{\mathcal{M}^r}$ and any given $e \in \mathcal{E}(m^r)$,

$$
\begin{aligned}
|\{m \mid e \in \mathcal{E}(m^r * m)\}| &= |\{m \in e(\mathcal{S}) \mid m \neq m_i, 1 \leq i \leq r\}| \\
&= k - r.
\end{aligned}
$$

Hence,

$$
\sum_{m \in \mathcal{M}(m^r)} \sum_{e \in \mathcal{E}(m' * m)} p(e) = (k - r) \sum_{e \in \mathcal{E}(m^r)} p(e),
$$

thus, (iii) follows from (ii).

Finally, it is trivial that $\overline{\mathcal{M}^1} = v$. When $r \geq 2$ for each $m^r \in \overline{\mathcal{M}^r}$,

$$
|\{m^{r-1} \mid m^{r-1} \subset m^r\}| = r.
$$

By (iii) we have

$$
|\overline{\mathcal{M}^r}| = \frac{1}{r} \sum_{m^{r-1} \in \overline{\mathcal{M}^{r-1}}} |\mathcal{M}(m^{r-1})| = \frac{(k - r + 1)|\overline{\mathcal{M}^{r-1}}|}{rP_{r-1}}.
$$

Thus, (iv) follows by induction on r. ⬜

We say that the probability distribution p_{S^r} is message uniform if the property (i) in Corollary 3.2 holds.

We write $\mathcal{M}(e)$ instead of $e(\mathcal{S})$ for each $e \in \mathcal{E}$ in the following; it is a subset of $\mathcal{M}$ with k elements. Thus, we have a family of k-subsets

$$\{\mathcal{M}(e) \subset \mathcal{M} \mid e \in \mathcal{E}\}. \tag{3.11}$$

A subset is also called a block in the combinatorial design theory. We will see that the successful probability of spoofing attack P_r is determined by the distribution of those blocks of Equation (3.11) in $\mathcal{M}$ at great extent. For a perfect authentication scheme, this family of blocks must have some special properties. We have to introduce some concepts of combinatorial designs first.

DEFINITION 3.3   Let v, b, k, λ, t be positive integers with $t \leq k$. A *partially balanced t-design* (**PBD**) $t - (v, b, k; \lambda, 0)$ is a pair $(\mathcal{N}, \mathcal{F})$ where $\mathcal{N}$ is a set of v points and $\mathcal{F}$ is a family of b subsets of $\mathcal{N}$, each of cardinality k (called blocks) such that any t-subset of $\mathcal{N}$ either occurs together in exactly λ blocks or does not occur in any block.

The structure in Definition 3.3 is also known as packing when $\lambda = 1$.

DEFINITION 3.4   If a *partially balanced t-design* $t - (v, b, k; \lambda_t, 0)$ is a *partially balanced r-design* $r - (v, b, k; \lambda_r, 0)$ for $1 \leq r < t$ as well, then it is called a *strong partially balanced t-design* (**SPBD**) and is denoted by $t - (v, b, k; \lambda_1, \lambda_2, \cdots, \lambda_t, 0)$.

DEFINITION 3.5   Let v, k, λ, t be positive integers, $t \leq k$. Suppose $\mathcal{M}$ is a set of v points and $\mathcal{E}$ is a family of k-subsets (blocks) of $\mathcal{M}$. A pair $(\mathcal{M}, \mathcal{E})$ is called a *t-design* if any t-subset of $\mathcal{M}$ occurs exactly in λ blocks. Such a t-design is denoted by $t - (v, k, \lambda)$.

It is easy to see that a t-design is a special case of a partially balanced t-design defined in Definition 3.3. The t-designs have already been extensively studied in the theory of block designs. The concept of partially balanced t-designs is a generalization of the concept of t-designs. A t-design is always strong. Precisely, a $t - (v, k, \lambda)$ design is a $r - (v, k, \lambda_r)$ design, where $1 \leq r < t$ and

$$\lambda_r = \frac{(v - r)(v - r - 1) \cdots (v - t + 1)}{(k - r)(k - r - 1) \cdots (k - t + 1)}. \tag{3.12}$$

Usually, we assume that any point of $\mathcal{M}$ appears in at least one block $\mathcal{M}(e)$, otherwise the point can be dismissed from $\mathcal{M}$. So we only concern the SPBD which is also a 1-design.

THEOREM 3.1
An authentication scheme $\mathcal{A} = (\mathcal{S}, \mathcal{M}, \mathcal{E}, p_S, p_E)$ is t-fold perfect if and only

if the pair

$$(\mathscr{M}, \{\mathscr{M}(e) \mid e \in \mathscr{E}\}) \tag{3.13}$$

is a SPBD $t - (v, b, k; \lambda_1, \lambda_2, \cdots, \lambda_t, 0)$ *with* $\lambda_t = 1$, p_E *is uniform, and* p_{S^r}, $1 \le r \le t$ *are message uniform (p_{S^t} is always message uniform when* $\lambda_t = 1$). *Where*

$$v = |\mathscr{M}|, \quad b = |\mathscr{E}|, \quad k = |\mathscr{S}|,$$

$$\lambda_r = (P_r P_{r+1} \cdots P_{t-1})^{-1}, \ 1 \le r \le t-1.$$

PROOF   Assume that $\mathscr{A}$ is t-fold perfect, then $\mathscr{A}$ is t-fold key-entropy minimal, p_E is uniform, and $H(E|M^t) = 0$ by Corollary 3.1. The last equality means that

$$-\sum_{m^t} \sum_{e \in \mathscr{E}(m^t)} p(e, m^t) \log p(e|m^t) = 0.$$

Each term in this sum is nonnegative; thus, for any $m^t \in \mathscr{M}^t$, if $m^t \notin \overline{\mathscr{M}^t}$ then $|\mathscr{E}(m^t)| = 0$; if $m^t \in \overline{\mathscr{M}^t}$ then $p(e|m^t) = 1$ for all $e \in \mathscr{E}(m^t)$. This is only possible when $|\mathscr{E}(m^t)| = 1$. Hence, we know that $\lambda_t = 1$.

Since p_E is uniform, it follows from (ii) of Corollary 3.2 that

$$P_r = \frac{|\mathscr{E}(m^r * m)|}{|\mathscr{E}(m^r)|}, \quad 0 \le r \le t-1 \tag{3.14}$$

for any $m^r * m \in \overline{\mathscr{M}^{r+1}}$. (Note that $\mathscr{E}(m^0) = \mathscr{E}$.)

For any $m^r \in \mathscr{M}^r, 1 \le r \le t-1$, we have $|\mathscr{E}(m^r)| = 0$, or $m^r \in \overline{\mathscr{M}^r}$. In the latter case there exists $m^t \in \overline{\mathscr{M}^t}$ such that $m^r \subset m^t$. It follows from Equation (3.14) that

$$|\mathscr{E}(m^r)| = (P_r P_{r+1} \cdots P_{t-1})^{-1} (= \lambda_r), \quad 1 \le r \le t-1.$$

So far we have proved that the pair in Equation (3.13) is a SPBD with the given parameters. The condition that $S^r, 1 \le r \le t$ are message uniform is nothing but the property (i) of Corollary 3.2. Thus, the conditions given in the theorem are necessary.

Now we show that the conditions are also sufficient. Assume that the conditions hold. For any $m^r * m \in \overline{\mathscr{M}^{r+1}}, 0 \le r \le t-1$, and $e \in \mathscr{E}(m^r * m)$, we have

$$\begin{aligned}
\frac{p(e \mid m^r)}{p(e \mid m^r * m)} &= \frac{p(e, m^r) p(m^r * m)}{p(e, m^r * m) p(m^r)} \\
&= \frac{p(f_e(m^r)) \sum_{e' \in \mathscr{E}(m^r * m)} p(e') p(f_{e'}(m^r * m))}{p(f_e(m^r * m)) \sum_{e' \in \mathscr{E}(m^r)} p(e') p(f_{e'}(m^r))} \\
&= \frac{|\mathscr{E}(m^r * m)|}{|\mathscr{E}(m^r)|} = \frac{\lambda_{r+1}}{\lambda_r},
\end{aligned}$$

where $\lambda_0 = b$. The above ratios are constant, hence, $P_r, 0 \le r \le t-1$ achieve their information-theoretic bounds given in Proposition 3.1 and

$$P_r = \frac{\lambda_{r+1}}{\lambda_r}, \qquad 0 \le r \le t-1.$$

It is obvious that

$$b = \lambda_0 = (P_0 P_1 \cdots P_{t-1})^{-1}.$$

The theorem is proved.   □

Given an authentication code $(\mathscr{S}, \mathscr{M}, \mathscr{E})$ we have the pair in Equation (3.13). Conversely, a pair $(\mathscr{M}, \mathscr{F})$ where $\mathscr{F}$ is a family of k-blocks of $\mathscr{M}$, together with a set $\mathscr{S}$ of k elements, gives an authentication code. For each block $B \in \mathscr{F}$, let $e_B : \mathscr{S} \to \mathscr{M}$ be an injective mapping with $e_B(\mathscr{S}) = B$; and put $\mathscr{E} = \{e_B | B \in \mathscr{F}\}$. Then $(\mathscr{S}, \mathscr{M}, \mathscr{E})$ becomes an authentication code with $(\mathscr{M}, \mathscr{F})$ as its corresponding pair. Here we do not require that any two blocks B_1 and B_2 in $\mathscr{F}$ are distinct, but we require that the mappings e_{B_1} and e_{B_2} are distinct if $B_1 = B_2$. Theorem 3.1 shows that any t-fold perfect authentication code can be constructed by a SPBD $t - (v, b, k; \lambda_1, \cdots, \lambda_t, 0)$ with $\lambda_t = 1$. Any block in $\mathscr{F}$ appears exactly one time when $\lambda_t = 1$.

Suppose that $(\mathscr{M}, \mathscr{F})$ is a SPBD $t - (v, b, k; \lambda_1, \cdots, \lambda_t, 0)$ and r is a positive integer. Letting each block in $\mathscr{F}$ appear r times, we form a new k-block family $\mathscr{F}' = r\mathscr{F}$. Then $(\mathscr{M}, \mathscr{F}')$ is a new SPBD $t - (v, b', k; \lambda_1', \cdots, \lambda_t', 0)$ where $b' = rb$, $\lambda_i' = r\lambda_i$, for $1 \le i \le t$. Obviously, $\lambda_{i+1}'/\lambda_i' = \lambda_{i+1}/\lambda_i$, $0 \le i \le t-1$, where $\lambda_0' = b'$, $\lambda_0 = b$. Hence, a new authentication code can be constructed by $(\mathscr{M}, \mathscr{F}')$ which has the same parameters P_r, $0 \le r \le t-1$ with those of the constructed code by $(\mathscr{M}, \mathscr{F})$. This idea will be used to construct schemes with key-entropy minimal as well as with some perfect secrecy property (see Chapter 9).

Theorem 3.1 shows that when we study a perfect authentication scheme $\mathscr{A} = (\mathscr{S}, \mathscr{M}, \mathscr{E}, p_S, p_E)$, we only need to concern the combinatorial structure of the pair in Equation (3.13) determined by the corresponding authentication code $(\mathscr{S}, \mathscr{M}, \mathscr{E})$. Hence, we also call the A-code $(\mathscr{S}, \mathscr{M}, \mathscr{E})$ as perfect.

3.4   Perfect Cartesian Codes

For any source state $s \in \mathscr{S}$, put

$$\mathscr{M}(s) = \{m \in \mathscr{M} \mid m = e(s) \ for \ some \ e \in \mathscr{E}\},$$

this is the set of the messages which can be used to transmit the source state s. If for any two source states s_1 and s_2 the sets $\mathscr{M}(s_1)$ and $\mathscr{M}(s_2)$ are disjointed, this kind of authentication codes are called Cartesian codes. When a Cartesian code is used, once the transmitted message is observed one may know its corresponding conveyed source state. This means that the Cartesian code has no secrecy.

In this section we assume that $(\mathscr{S}, \mathscr{M}, \mathscr{E})$ is a t-fold perfect Cartesian code, and study its characterization. If $m^r \in \mathscr{M}^r$, $e \in \mathscr{E}(m^r)$, then $f_e(m^r) \in \mathscr{S}^r$ does not depend on e. Hence, the random variable $\mathscr{S}^r$ is message uniform, obviously.

COROLLARY 3.3
In a perfect Cartesian code, we have $|\mathscr{M}(s)| = P_0^{-1}$ for all source states s.

PROOF Let s be any source state, by (ii) of Corollary 3.2,

$$|\mathscr{M}(s)| \cdot P_0 = \sum_{m \in \mathscr{M}(s)} \sum_{e \in \mathscr{E}(m)} p(e) = \sum_{e \in \mathscr{E}} p(e) = 1,$$

hence

$$|\mathscr{M}(s)| = P_0^{-1},$$

which does not depend on s. □

Denote $P_0^{-1} = n$ which is a positive integer, hence, $v = nk$ for perfect Cartesian codes. Let $\mathscr{S} = \{s_1, s_2, \cdots, s_k\}$ and $s_{i_1}, s_{i_2}, \cdots, s_{i_t}, (t \leq k)$ be any t-subset of $\mathscr{S}$. Put

$$\mathscr{M}_{i_1, \cdots, i_t} = \{m^t = (m_1, \cdots, m_t) \mid m_r \in \mathscr{M}(s_{i_r}), 1 \leq r \leq t\}$$

We have $|\mathscr{M}_{i_1, \cdots, i_t}| = n^t$ by Corollary 3.3. For any $m^t \in \mathscr{M}_{i_1, \cdots, i_t}$ we have $|\mathscr{E}(m^t)| \leq 1$ by Theorem 3.1.

COROLLARY 3.4
In a t-fold perfect Cartesian code, let $s_{i_1}, \cdots, s_{i_t}$ be a t-subset of $\mathscr{S}$ and suppose that $|\mathscr{E}(m^t)| = 1$ for all $m^t \in \mathscr{M}_{i_1, \cdots, i_t}$. Let $s_{j_1}, \cdots, s_{j_t}$ be any t-subset of $\mathscr{S}$. Then

(i) $|\mathscr{E}| = n^t$;

(ii) $|\mathscr{E}(m^t)| = 1$ for all $m^t \in \mathscr{M}_{j_1, \cdots, j_t}$;

(iii) $|\mathscr{E}(m^r)| = n^{t-r}$ for all $m^r \in \mathscr{M}_{j_1, \cdots, j_r} (1 \leq r \leq t)$;

(iv) $P_0 = P_1 = \cdots = P_{t-1} = n^{-1}$.

PROOF We may assume that $\{i_1, \cdots, i_t\} = \{1, \cdots, t\}$. We have $|\mathscr{E}| \geq |\mathscr{M}_{1, \cdots, t}| = n^t$ by the assumption of the corollary. On the other hand, each encoding rule $e \in \mathscr{E}$ corresponds to a t-subset $m_e^t = \{e(s_1), \cdots, e(s_t)\} \in \mathscr{M}_{1,\cdots,t}$. Thus, (i) is proved. Any $e \in \mathscr{E}$ corresponds to a t-subset $e^t = \{e(s_{j_1}), \cdots, e(s_{j_t})\} \in \mathscr{M}_{j_1,\cdots,j_t}$, hence, (ii) follows according to that $|\mathscr{E}| = |\mathscr{M}_{j_1,\cdots,j_t}|$ and $\lambda_t = 1$. For any $m^r \in \mathscr{M}_{j_1,\cdots,j_r}$ we take $m_{j_p} \in \mathscr{M}(s_{j_p})(r < p \le t)$, thus, $m^t = \{m^r, m_{j_{r+1}}, \cdots, m_t\} \in \mathscr{M}_{j_1,\cdots,j_t}$ which corresponds to a unique encoding rule $e \in \mathscr{E}(m^t) \subset \mathscr{E}(m^r)$. There are n^{t-r} choices for $m_{j_p} \in \mathscr{M}(s_{j_p})(r < p \le t)$, thus, $n^{t-r} \le |\mathscr{E}(m^r)|$. On the other hand, each encoding rule $e \in \mathscr{E}(m^r)$ corresponds to a t-subset $m_e^t \in \mathscr{M}_{j_1,\cdots,j_t}$ such that $m^r \subset m_e^t \subset \mathscr{M}(e)$. It implies that $|\mathscr{E}(m^r)| \le n^{t-r}$. Therefore, (iii) is proved. Property (iv) is deduced from (iii) and Equation (3.14). □

We call a perfect Cartesian code as of Type I if it satisfies the assumption in Corollary 3.4, otherwise it is called of Type II. Now we look at some examples of perfect Cartesian codes.

Scheme 1

Let $\mathbb{F}_q$ be the finite field with q elements, and t be a positive integer less than q. Take $\mathscr{S} = \mathbb{F}_q, \mathscr{M} = AG(2, \mathbb{F}_q)$ (affine plane over $\mathbb{F}_q$), and define blocks

$$B(a_{t-1}, a_{t-2}, \cdots, a_0) = \{(x,y) : y = \sum_{i=0}^{t-1} a_i x^i, x \in \mathbb{F}_q\}$$

for each t-tuple $(a_{t-1}, a_{t-2}, \cdots, a_0), a_i \in \mathbb{F}_q$. Let $\mathscr{B}$ be the family of all blocks. There are totally q^t such blocks. For each block $B(a_{t-1}, \cdots, a_0)$ we define an encoding rule e such that

$$e(x) = \sum_{i=0}^{t-1} a_i x^i, \forall x \in \mathscr{S}$$

Thus, we have $|\mathscr{E}| = q^t$. We can see that $|\mathscr{M}(x)| = q$ for all $x \in \mathscr{S}$. Take t distinct elements x_i $(1 \le i \le t)$ of $\mathbb{F}_q$. For any r $(1 \le r \le t)$ points $m^r = \{(x_1, y_1), (x_2, y_2), \cdots, (x_r, y_r)\}$, if they are in a block $B(a_{t-1}, a_{t-2}, \cdots, a_0)$, then they satisfy

$$a_{t-1}x_1^{t-1} + a_{t-2}x_1^{t-2} + \cdots + a_1 x_1 + a_0 = y_1$$
$$a_{t-1}x_2^{t-1} + a_{t-2}x_2^{t-2} + \cdots + a_1 x_2 + a_0 = y_2$$
$$\cdots\cdots\cdots\cdots\cdots$$
$$a_{t-1}x_r^{t-1} + a_{t-2}x_r^{t-2} + \cdots + a_1 x_r + a_0 = y_r$$

This is a system of linear equations with unknown $a_{t-1}, a_{t-2}, \cdots, a_0$. The rank of the coefficient matrix is r, it has q^{t-r} solutions $(a_{t-1}, a_{t-2}, \cdots, a_0)$. This means that $|\mathscr{E}(m^r)| = q^{t-r}$. Therefore $(\mathscr{M}, \mathscr{B})$ is a $t - (q^2, q^t, q; \lambda_1, \lambda_2, \cdots, \lambda_t, 0)$ design with $\lambda_r = q^{t-r}$, $1 \le r \le t$. This is a t-fold perfect Cartesian code. If $r = t$ the above system of linear equations always has a unique solution, and hence, this code is of Type I.

Scheme 2

If we require that $a_{t-1} \neq 0$ in Scheme 1, then we have a new code with $q^{t-1}(q-1)$ blocks defined as

$$B'(a_{t-1}, a_{t-2}, \cdots, a_0) = \{(x,y): \ y = \sum_{i=0}^{t-1} a_i x^i, x \in \mathbb{F}_q\}$$

for each t-tuple $(a_{t-1}, a_{t-2}, \cdots, a_0), a_i \in \mathbb{F}_q, a_{t-1} \neq 0$. Let $x_i(1 \leq i \leq t)$ be distinct elements of $\mathbb{F}_q$ as above. The new code has $|\mathscr{E}| = q^{t-1}(q-1)$ and $|\mathscr{E}(m^r)| = q^{t-1-r}(q-1)$ for all $m^r = \{(x_1, y_1), \cdots, (x_r, y_r)\}, (1 \leq r \leq t-1)$. For t-subset $m^t = \{(x_1, y_1), \cdots, (x_t, y_t)\}$, put

$$D = \begin{vmatrix} y_1 & x_1^{t-2} & x_1^{t-3} & \cdots & x_1 & 1 \\ y_2 & x_2^{t-2} & x_2^{t-3} & \cdots & x_2 & 1 \\ \vdots & & & \ddots & & \\ y_t & x_t^{t-2} & x_t^{t-2} & \cdots & x_t & 1 \end{vmatrix}$$

When $D = 0$, m^t is not contained in any block while for $D \neq 0$, m^t is contained in a unique block. Therefore, the new code is a t-fold perfect Cartesian code of Type II.

Before giving the next scheme we introduce some knowledge of n-dimensional projective space $PG(n, \mathbb{F}_q)$ over the finite field $\mathbb{F}_q$, where n is an arbitrary positive integer. Each point of $PG(n, \mathbb{F}_q)$ is represented by its homogeneous coordinate

$$\lambda(x_0, x_1, \cdots x_n)$$

where $x_i \in \mathbb{F}_q$ $(0 \leq i \leq n)$ are not all zero and λ is an arbitrary nonzero element of $\mathbb{F}_q$. The set of all points which satisfy a system of linear independent homogeneous equations

$$a_{10}X_0 + a_{11}X_1 + \cdots + a_{1n}X_n = 0$$
$$a_{20}X_0 + a_{21}X_1 + \cdots + a_{2n}X_n = 0$$
$$\cdots\cdots\cdots$$
$$a_{r0}X_0 + a_{r1}X_1 + \cdots + a_{rn}X_n = 0 \qquad (3.15)$$

where $1 \leq r \leq n$, is called an $(n-r)$-dimensional subspace of $PG(n, \mathbb{F}_q)$. Assuming that the $n-r+1$ vectors

$$p_i = (x_{i0}, x_{i1}, \cdots, x_{in}), \quad 0 \leq i \leq n-r \qquad (3.16)$$

form a basis of the solutions of Equation (3.15), then any point of $PG(n, \mathbb{F}_q)$ which satisfies Equation (3.15) can be represented by

$$x_0 p_0 + x_1 p_1 + \cdots + x_{n-r} p_{n-r} \qquad (3.17)$$

where $x_i \in \mathbb{F}_q$, $(0 \le i \le n-r)$ are not all zero. Hence, the $(n-r)$-dimensional subspace in Equation (3.17) can be regarded as the space $PG(n-r, \mathbb{F}_q)$. Conversely, any set of $t+1$ $(t = n-r)$ linearly independent points $p_i = (x_{i0}, \cdots, x_{in})$ $(0 \le i \le t)$ of $PG(n, \mathbb{F}_q)$ determines a t-dimensional subspace in Equation (3.17) and the system of equations, which defines the subspace, can be constructed by taking any basis of the solutions for the system of equations

$$
\begin{aligned}
a_0 x_{00} + a_1 x_{01} + \cdots + a_n x_{0n} &= 0 \\
a_0 x_{10} + a_1 x_{11} + \cdots + a_n x_{1n} &= 0 \\
\cdots\cdots\cdots &= 0 \\
a_0 x_{t0} + a_1 x_{t1} + \cdots + a_n x_{tn} &= 0.
\end{aligned}
$$

as its coefficient matrix.

Let $N(n,t)$ $(0 \le t < n)$ denote the number of t-dimensional subspaces in $PG(n, \mathbb{F}_q)$.

LEMMA 3.5

$$
N(n,t) = \frac{(q^{n+1} - 1)(q^n - 1)\cdots(q^{t+2} - 1)}{(q^{n-t} - 1)(q^{n-t-1} - 1)\cdots(q - 1)}.
$$

PROOF   Each t-dimensional subspace is defined by a system of linear homogeneous equations of order $n-t$. Its coefficient matrix $A = A_{(n-t)\times(n+1)}$ is of rank $n-t$. The number of all full rank $(n-t) \times (n+1)$ matrices is

$$
(q^{n+1} - 1)(q^{n+1} - q)\cdots(q^{n+1} - q^{n-t-1}).
$$

But two matrices A and $A' = A'_{(n-t)\times(n+1)}$ define the same t-dimensional subspace if and only if there exists a nonsingular matrix S of order $n-t$ such that

$$
A = SA'.
$$

The number of all nonsingular matrices of order $n - t$ is

$$
(q^{n-t} - 1)(q^{n-t} - q)\cdots(q^{n-t} - q^{n-t-1}),
$$

hence

$$
\begin{aligned}
N(n,t) &= \frac{(q^{n+1} - 1)(q^{n+1} - q)\cdots(q^{n+1} - q^{n-t-1})}{(q^{n-t} - 1)(q^{n-t} - q)\cdots(q^{n-t} - q^{n-t-1})} \\
&= \frac{(q^{n+1} - 1)(q^n - 1)\cdots(q^{t+2} - 1)}{(q^{n-t} - 1)(q^{n-t-1} - 1)\cdots(q - 1)}.
\end{aligned}
$$

□

Let $t < r < n$. Since an r-dimensional subspace in $PG(n, \mathbb{F}_q)$ can be regarded as $PG(r, \mathbb{F}_q)$, we have

COROLLARY 3.5

Suppose $0 \leq t < r$. The number of t-dimensional subspaces contained in an r-dimensional subspace of $PG(n, \mathbb{F}_q)$ is

$$N(r, t) = \frac{(q^{r+1} - 1)(q^r - 1) \cdots (q^{t+2} - 1)}{(q^{r-t} - 1)(q^{r-t-1} - 1) \cdots (q - 1)}.$$

{Note that $N(r, t)$ does not depend on n.}

LEMMA 3.6

Suppose that $0 \leq t < r < n$. Let L be a fixed t-dimensional subspace of $PG(n, \mathbb{F}_q)$. The number of r-dimensional subspaces containing L is

$$Q_n(t, r) = \frac{(q^{n-t} - 1) \cdots (q^{n-r+1} - 1)}{(q^{r-t} - 1) \cdots (q - 1)}.$$

PROOF We count the number D of the pairs (H_r, H_t), where H_r is an r-dimensional subspace and H_t is a t-dimensional subspace contained in H_r, by two ways. There are $N(n, r)$ choices of H_r in $PG(n, \mathbb{F}_q)$, and for each fixed H_r there are $N(r, t)$ choices of H_t contained in the H_r, hence

$$D = N(n, r) \cdot N(r, t).$$

On the other hand, there are $N(n, t)$ choices of H_t in $PG(n, \mathbb{F}_q)$, and for each fixed H_t there are $Q_n(t, r)$ choices of H_r containing H_t, hence,

$$D = N(n, t) \cdot Q_n(t, r).$$

Therefore,

$$Q_n(t, r) = \frac{N(n, r) N(r, t)}{N(n, t)}$$

$$= \frac{(q^{n-t} - 1) \cdots (q^{n-r+1} - 1)}{(q^{r-t} - 1) \cdots (q - 1)}.$$

$\square$

Let us look at the projective plane $PG(2, \mathbb{F}_q)$ as an example. There are $N(2, 0) = q^2 + q + 1$ points (0-dimensional subspaces) and $N(2, 1) = q^2 + q + 1$ lines (1-dimensional subspaces) in $PG(2, \mathbb{F}_q)$ by Lemma 3.5, and there are $N(1, 0) = q + 1$ points on a line by Corollary 3.5. There are $Q_2(0, 1) = q + 1$ lines passing through one fixed point in $PG(2, \mathbb{F}_q)$ by Lemma 3.6.

LEMMA 3.7

In the space $PG(2, \mathbb{F}_q)$ any two different lines always intersect at one point.

PROOF   Let $aX_0 + bX_1 + cX_2 = 0$ and $a'X_0 + b'X_1 + c'X_2 = 0$ be different lines. Their intersect points are determined by the solutions of the system of linear homogeneous equations

$$aX_0 + bX_1 + cX_2 = 0$$
$$a'X_0 + b'X_1 + c'X_2 = 0.$$

Since the rank of the coefficient matrix

$$\begin{pmatrix} a & b & c \\ a' & b' & c' \end{pmatrix}$$

is 2, the lemma is proved.   □

Scheme 3

Fix a line L in $PG(2, \mathbb{F})$. The points on L are regarded as the source states, the points not lying on L are regarded as the encoding rules, and the lines different from L are regarded as the messages. Thus we have $|\mathscr{S}| = q+1, |\mathscr{E}| = (q^2+q+1) - (q+1) = q^2$, and $|\mathscr{M}| = (q^2+q+1) - 1 = q^2+q$. Given a source state s and an encoding rule e, there is a unique line m passing through s and e. The source state s is encoded into the message m by the encoding rule e. We can see that a message m is valid under the encoding rule e if the point e is located on the line m. Any line m, different from L, contains q points not lying on L, hence $|\mathscr{E}(m)| = q$. Any two lines m_1, m_2 intersect at one point by Lemma 3.7; this means that $|\mathscr{E}(m_1, m_2)| = 1$ which implies that the assumption of Corollary 3.4 is satisfied. If we assume that $\mathscr{E}$ has uniform distribution, this code becomes a 2-fold perfect Cartesian code of Type I with $P_0 = P_1 = 1/q$.

Scheme 4

This scheme is a generalization of Scheme 3. Fix an r-dimensional subspace L in $PG(n, \mathbb{F}_q)$ where $1 \leq r \leq n-1$. All t-dimensional subspaces $(0 \leq t < r)$ contained in L are regarded as source states. All $(n - r - 1)$-dimensional subspaces which have an empty intersection with L are regarded as encoding rules. All $(n - r + t)$-dimensional subspaces intersecting L in a t-dimensional subspace are regarded as messages. The source state s is encoded by the encoding rule e into the message $< s, e >$, which is an $(n - r + t)$-dimensional subspace generated by s and e. (By the dimensional formula: $dim < s, e > = dim(s) + dim(e) - dim(s \cap e) = t + (n - r - 1) - (-1) = n - r + t$.)

Any message m {a $(n - r + t)$-dimensional subspace} can be only used to transmit a unique source state $s = m \cap L$. Hence, Scheme 4 is Cartesian. The number of source states is $N(r, t)$ by Corollary 3.5.

Let $A = A_{(n-r)\times(n+1)}$ be the coefficient matrix of the equation system which defines the subspace L, and let $B = B_{(r+1)\times(n+1)}$ be the coefficient matrix of an equation system which defines an $(n-r-1)$-dimensional subspace having an empty intersection with L. Then the matrix of order $n+1$

$$\begin{pmatrix} A \\ B \end{pmatrix}$$

is nonsingular. The matrices B and $B'_{(r+1)\times(n+1)}$ define the same $(n-t-1)$-dimensional subspace if and only if there exists a nonsingular matrix S of order $r+1$ such that $B = SB'$. Hence, the number of encoding rules of Scheme 4 is

$$\frac{(q^{n+1} - q^{n-r})(q^{n+1} - q^{n-r-1}) \cdots (q^{n+1} - q^{n})}{(q^{r+1} - 1)(q^{r+1} - q) \cdots (q^{r+1} - q^r)} = q^{(n-r)(r+1)}.$$

Let $C = C_{(r-t)\times(n+1)}$ be the coefficient matrix of any equation system which defines an $(n-r+t)$-dimensional subspace intersecting L in a t-dimensional subspace. Then the matrix

$$\begin{pmatrix} A \\ C \end{pmatrix}$$

of $(n-t) \times (n+1)$ is of full rank. Similarly, the matrix C and $C'_{(n-t)\times(n+1)}$ define the same $(n-r+t)$-dimensional subspace if and only if there exists a nonsingular matrix S of order $r-t$ such that $C = SC'$. Hence, the number of messages is

$$\frac{(q^{n+1} - q^{n-r})(q^{n+1} - q^{n-r+1}) \cdots (q^{n+1} - q^{n-t+1})}{(q^{r-t} - 1)(q^{r-t} - q) \cdots (q^{r-t} - q^{r-t-1})} = q^{(n-r)(r-t)} N(r,t).$$

Given a message $m = <s,e>$, suppose that

$$m = <p_0, p_1, \cdots, p_{n-r+t}>$$

where $p_i \in PG(n, \mathbb{F}_q), 0 \le i \le n-r+t$ are linearly independent, and

$$s = <p_0, p_1, \cdots, p_t>,$$
$$e = <p_{t+1}, \cdots, p_{n-r+t}> .$$

Assume that $q_i \in s, (t+1 \le i \le n-r+t)$, put

$$e^* = <p_{t+1} + q_{t+1}, \cdots, p_{n-r+t} + q_{n-r+t}> . \qquad (3.18)$$

It is easy to show that

$$<s,e> = <s, e^*>,$$

e^* is also an encoding rule, i.e., e^* is an $(n-r-1)$-dimensional subspace and $e^* \cap L = \emptyset$. Furthermore, we have $e \neq e^*$ when $q_i(t+1 \leq i \leq n-r+t)$ are not all zero, otherwise if $e = e^*$, say $q_{t+1} \neq 0$, then there exists c_i such that

$$p_{t+1} + q_{t+1} = \sum_{i=t+1}^{n-r+t} c_i p_i,$$

which implies that $q_{t+1} \in s \cap e$; this is a contradiction with the assumption that $e \cap L = \emptyset$. Similarly, we can show that any different choice of the set $(q_i,\ t+1 \leq i \leq n-r+t)$ corresponds to a different encoding rule, hence,

$$|\mathscr{E}(m)| \geq q^{(n-r)(t+1)}, \tag{3.19}$$

for any message m. On the other hand, suppose that an encoding rule

$$e' = < p'_{t+1}, \cdots, p'_{n-r+t} > \in \mathscr{E}(m),$$

then $m = < s, e > = < s, e' >$ and there exists a matrix $\{\alpha_{ij}\}$ such that

$$p'_i = \sum_{j=t+1}^{n-r+t} \alpha_{ij} p_j + q'_i \quad (t+1 \leq i \leq n-r+t),$$

where $q'_i \in s$. The matrix $(\alpha_{ij})_{t+1 \leq i,j \leq n-r+t}$ is nonsingular, otherwise there exist $c_i(t+1 \leq i \leq n-r+t)$, which are not all zero, such that

$$\sum_{i=t+1}^{n-r+t} c_i p'_i \in s \subset L,$$

which is also a contradiction. Therefore the encoding rule e' can be represented with the form

$$e' = < p_{t+1} + q_{t+1}, \cdots, p_{n-r+t} + q_{n-r+t} >,$$

where $q_i \in s, (t+1 \leq i \leq n-r+t)$. In other words, any encoding rule of $\mathscr{E}(m)$ can be represented in the form of Equation (3.18). Hence, the equality in Equation (3.19) holds. Thus,

$$P_0 = \frac{|\mathscr{E}(m)|}{|\mathscr{E}|} = q^{-(n-r)(r-t)},$$

when the encoding rules have a uniform probability distribution.

Given any two different messages m and m', if there exists an encoding rule $e \in \mathscr{E}(m, m')$, then $m = < s, e >, m' = < s', e >$, s and s' are two different source states. Let

$$m = < p_0, \cdots, p_t, p_{t+1}, \cdots, p_{n-r+t} >,$$
$$m' = < p'_0, \cdots, p'_t, p_{t+1}, \cdots, p_{n-r+t} > .$$

Then any encoding rule e^* in $\mathscr{E}(m, m')$ has the form

$$e^* =< p_{t+1} + q_{t+1}, \cdots, p_{n-r+t} + q_{n-r+t} >,$$

where $q_i \in s \cap s'$, $(t + 1 \leq i \leq n - r + t)$. The maximal dimension of the projective subspace $s \cap s'$ is $t - 1$, hence the maximal value of $|\mathscr{E}(m, m')|$ is $q^{(n-r)t}$, and

$$P_1 = \frac{q^{(n-r)t}}{q^{(n-r)(t+1)}} = q^{-(n-r)}.$$

Each Cartesian code can be represented by a $b \times k$ matrix where the rows are indexed by encoding rules, the columns are indexed by source states, the entry in row e and column s is $e(s)$, and where the messages in each set $\mathscr{M}(s)$ are labelled by the symbols $1, 2, \cdots, n$. This matrix is called the encoding matrix of the code.

DEFINITION 3.6   *An $N \times k$ array A with entries from n symbols is said to be an orthogonal array with n levels, strength t $(0 \leq t \leq k)$, and index λ, if every $N \times t$ subarray of A contains each t-tuple of the symbols exactly λ times as a row.*

It is easy to see that $\lambda = N/n^t$. We will denote such an array by $OA(N, k, n, t)$. It is customary to say that the orthogonal array has index unity when $\lambda = 1$.

Regard each column of an orthogonal array $OA(N, k, n, t)$ as a source state, each row as an encoding rule, and regard symbols in different columns as different messages, then it is easy to see that this orthogonal array is an $r - (kn, n^t, k; n^{t-r}, 0)$ design for $2 \leq r \leq t$ and a $1 - (kn, n^t, k, n^{t-1})$ design as well.

Based on Theorem 3.1 we have

THEOREM 3.2
A Cartesian code $(\mathscr{S}, \mathscr{M}, \mathscr{E})$ is t-fold perfect of Type I if and only if its encoding matrix is an orthogonal array $OA(n^t, k, n, t)$ of index unity and p_E is uniform where $k = |\mathscr{S}|$, $n = |\mathscr{M}(s)|$ for each $s \in \mathscr{S}$, and $|\mathscr{E}| = n^t$.

Theorem 3.1 reduces the construction of perfect authentication codes to the construction of SPBD. For the special case of Cartesian codes of Type I it is reduced to the construction of orthogonal arrays of index unity.

PROPOSITION 3.2
A orthogonal array $OA(n^t, k, n, t)$ is an $r - (kn, n^t, k; n^{t-r}, 0)$ design for $2 \leq r \leq t$ and is a $1 - (kn, n^t, k, n^{t-1})$ design as well.

3.5   Combinatorial Bound

In this section we discuss the combinatorial bound on P_r. Recall that $|\mathscr{S}| = k, |\mathscr{M}| = v$, and $|\mathscr{E}| = b$. Let $m^r = (m_1, m_2, \cdots, m_r) \in \overline{\mathscr{M}^r}$ {here m_i $(1 \le i \le r)$ is distinct by the Assumption 2.1}. Also, we know that only when $m \ne m_i$ for $1 \le i \le r$ it is possible for

$$P(m|m^r) = \sum_{e \in \mathscr{E}(m' * m)} p(e|m^r)$$

to be nonzero. So

$$
\begin{aligned}
(v-r) \cdot \max_m P(m|m^r) &\ge \sum_m P(m|m^r) = \sum_m \sum_{e \in \mathscr{E}(m' * m)} p(e|m^r) \\
&= \sum_{e \in \mathscr{E}(m')} p(e|m^r) \sum_{\substack{m \in \mathscr{M}(e) \\ m \ne m_i}} 1 \\
&= (k-r) \sum_{e \in \mathscr{E}(m')} p(e|m^r) \\
&= k - r,
\end{aligned}
\tag{3.20}
$$

therefore,

$$
\begin{aligned}
(v-r) \cdot P_r = (v-r) \sum_{m' \in \mathscr{M}'} p(m^r) \max_m P(m|m^r) \\
\ge (k-r) \sum_{m' \in \mathscr{M}'} p(m^r) = k - r
\end{aligned}
\tag{3.21}
$$

From Equations (3.20) and (3.21) we know that the equality in Equation (3.21) holds if and only if

$$P(m|m^r) = \sum_{e \in \mathscr{E}(m' * m)} p(e|m^r) = \frac{k-r}{v-r}$$

for any $m^r = (m_1, m_2, \cdots, m_r) \in \overline{\mathscr{M}^r}$ and any $m \in \mathscr{M}$ with $m \ne m_i, 1 \le i \le r$. Then the following proposition is proved.

PROPOSITION 3.3
We have

$$P_r \ge \frac{k-r}{v-r}$$

for $r = 0, 1, \cdots, k-1$. The equality holds if and only if

$$P(m|m^r) = \frac{k-r}{v-r}$$

is satisfied for any $m^r = (m_1, m_2, \cdots, m_r) \in \overline{\mathcal{M}^r}$ and any $m \in \mathcal{M}$ with $m \neq m_i, 1 \leq i \leq r$.

The bound given in Proposition 3.3 is called the combinatorial bound. Suppose that in an authentication scheme $\mathcal{A} = (\mathcal{S}, \mathcal{M}, \mathcal{E}, p_S, p_E)$, P_r achieves the combinatorial bound, for $0 \leq r \leq t - 1$, that is

$$P_r = \frac{k - r}{v - r}, \quad r = 0, 1, \cdots, t - 1,$$

for $t \leq k$. Then from Proposition 3.3 we have $m^r * m \in \overline{\mathcal{M}^{r+1}}$ for any $m^r \in \overline{\mathcal{M}^r}$, and any $m \neq m_i, 1 \leq i \leq r$. By the assumption of $\mathcal{E}(m) \neq \emptyset$ for any $m \in \mathcal{M}$, i.e., $\overline{\mathcal{M}} = \mathcal{M}$, then it is easy to show that $\overline{\mathcal{M}^r} = \mathcal{M}^r$ for $1 \leq r \leq t$. Hence, $|\mathcal{E}(m^t)| \geq 1$ for any $m^t \in \mathcal{M}^t$. There are a total of C_v^t t-subsets of $\mathcal{M}$ and C_k^t t-subsets of $\mathcal{M}$ in each block with k points. Therefore,

$$b \geq \frac{C_v^t}{C_k^t} = \frac{v(v-1)\cdots(v-t+1)}{k(k-1)\cdots(k-t+1)}$$

When the above equality holds, $|\mathcal{E}(m^t)| = 1$ holds for any $m^t \in \mathcal{M}^t$.

THEOREM 3.3

Let $(\mathcal{S}, \mathcal{M}, \mathcal{E}, p_S, p_E)$ be an authentication scheme. The following properties

$$P_r = \frac{k - r}{v - r}, \ 0 \leq r < t \leq k \tag{3.22}$$

$$b = \frac{v(v-1)\cdots(v-t+1)}{k(k-1)\cdots(k-t+1)} (= (P_0 \cdots P_{t-1})^{-1}) \tag{3.23}$$

are satisfied if and only if $(\mathcal{M}, \{\mathcal{M}(e)|e \in \mathcal{E}\})$ is a $t - (v, k, 1)$ design, the probability distribution on $\mathcal{E}$ is uniform and the probability distribution on $\mathcal{S}^r \ (1 \leq r \leq t)$ is message uniform.

PROOF First, we suppose the conditions in Equations (3.22) and (3.23) are satisfied. It reduces from Equation (3.23) that $(\mathcal{M}, \{\mathcal{M}(e)|e \in \mathcal{E}\})$ is a $t - (v, k, 1)$ design as shown above. By Theorem 3.1 we know that p_E is uniform and $p_{S^r} \ (1 \leq r \leq t)$ is message uniform.

Now we suppose that $(\mathcal{M}, \{\mathcal{M}(e)|e \in \mathcal{E}\})$ is a $t - (v, k, 1)$ design, p_E is uniform, and $p_{S^r} \ (1 \leq r \leq t)$ is message uniform. Then for any $r \ (0 \leq r \leq t - 1)$,

$$\sum_{e \in \mathcal{E}(m^r * m)} p(e|m^r) = \frac{\sum_{e \in \mathcal{E}(m^r * m)} p(e)p(f_e(m^r))}{\sum_{e \in \mathcal{E}(m^r)} p(e)p(f_e(m^r))}$$

$$= \frac{|\mathcal{E}(m^r * m)|}{|\mathcal{E}(m^r)|} = \frac{\lambda_{r+1}}{\lambda_r} = \frac{k - r}{v - r}.$$

From Proposition 3.3, P_r achieves the combinatorial bound for r ($0 \le r \le t - 1$). Furthermore,

$$b = \lambda_0 = (P_0 P_1 \cdots P_{t-1})^{-1}$$
$$= \frac{v(v-1)\cdots(v-t+1)}{k(k-1)\cdots(k-t+1)}$$

The theorem is proved.   ∎

Noting that a t-design is a partially balanced t-design, by comparing with Theorem 3.3 and Theorem 3.1 we obtain

CLAIM 3.1

In an authentication scheme, if P_r achieves its combinatorial bound for all $r, 0 \le r \le t - 1$, and b achieves its lower bound $(P_0 \cdots P_{t-1})^{-1}$, then P_r ($0 \le r \le t-1$) achieves its information-theoretic bound too. We call the authentication schemes which satisfy the conditions of the Theorem 3.3 as t-fold perfect authentication schemes with combinatorial bounds.

In fact we have

PROPOSITION 3.4

Suppose that $(\mathscr{S}, \mathscr{M}, \mathscr{E}, p_S, p_E)$ is a t-fold perfect A-code. Then

$$P_r = 2^{H(E|M^{r+1}) - H(E|M^r)} \ge \frac{k-r}{v-r}, \quad 0 \le r < t.$$

PROOF   The first equality is obvious. For any $m^{r+1} = m^r * m \in \overline{\mathscr{M}^{r+1}}$, we have

$$P_r = \frac{|\mathscr{E}(m^r * m)|}{|\mathscr{E}(m^r)|}, \quad 0 \le r < t$$

and

$$|\mathscr{M}(m^r)| = \frac{|\mathscr{E}(m^r)|(k-r)}{|\mathscr{E}(m^{r+1})|} \le v - r$$

by (ii) and (iii) of Corollary 3.2. The desired result follows.   ∎

3.6   Comments

It was Shannon [40] who first studied the theory of secrecy based on the information theory. This was a milestone in the development of cryptology. Simmons [41] developed an analogous theory for authentication.

The information-theoretic bound for P_0 was first proved by Simmons [41], then Massey [23] and Sgarro [38] provided a short proof of the result. Walker [51] proved the inequality in Equation (3.3) for Cartesian codes. The Proposition 3.1 is due to the author [28]. It proved the inequality in Equation (3.3) for general (Cartesian or non-Cartesian) authentication codes and provided a simple formulation in Equation (3.4) of the necessary and sufficient condition for the equality in Equation (3.3) to hold, which in turn leads to finding the combinatorial structure of perfect authentication codes given in Theorem 3.1. This result was presented at the rump section of Asiacrypt '91 [36]. Smeets [45], Sgarro [39] and Rosenbaun [35] all independently proved the inequality in Equation (3.3), but their formulations of the condition for the equality in Equation (3.3) are more complicated and not convenient for use.

Theorem 3.1 is due to the author. The text pp. 19-25 is cited from the text pp. 177-188 of [28] with kind permission of Springer Science and Business Media.

Scheme 2 of Section 3.4 was presented by Wang [54], and Scheme 3 was presented by Gilbert, MacWilliams, and Sloane [12], which was the first authentication code constructed.

The result of Theorem 3.3, when $t = 2$, was first mentioned by Schöbi [37] and De Soete [9].

3.7   Exercises

3.1 Prove Equation (3.12).

3.2 Prove Proposition 3.2.

3.3 If $P_0 = P_1 = \cdots = P_{t-1} = P$, prove that $P \geq |\mathscr{E}|^{-1/t}$.

3.4 Let k, t be two positive integers with $k \geq t$. Suppose there exists a set of k vectors $\alpha_i = (c_{i1}, c_{i2}, \cdots, c_{it}) \in \mathbb{F}_q^t, 1 \leq i \leq k$ and any t vectors in the set are linearly independent. Set $\mathscr{S} = \{1, 2, \cdots, k\}, \mathscr{M} = \mathbb{F}_q$, and $\mathscr{E} = \mathbb{F}_q^t$. For any source state i and encoding rule $e = (a_1, a_2, \cdots, a_t) \in \mathbb{F}_q^t$ define $e(i) = \sum_{j=1}^t c_{ij} a_j$. Prove that the code constructed is t-fold perfect with $P_0 = P_1 = \cdots = P_{t-1} = 1/q$ if the encoding rules have a uniform probability distribution.

3.5 Let $\mathbb{F}_q^2$ be the vector space of dimension 2 over the finite field $\mathbb{F}_q$. Fix a line L in $\mathbb{F}_q^2$. The points on L are regarded as source states, the points not on L are regarded as encoding rules, and the lines not parallel to L and equal to L are regarded as messages. Given a source state s and an encoding rule e, there is a unique line m passing through s and e, then we define $m = e(s)$. Prove that $|\mathscr{S}| = q, |\mathscr{E}| = q^2 - q$, and $\mathscr{M} = q^2$, and that this is a 2-fold perfect Cartesian code of Type II with $P_0 = 1/q$ and $P_1 = 1/(q-1)$ if the encoding rules have a uniform probability distribution.

3.6 Fix a line L and a point P on L in $PG(2, \mathbb{F}_q)$. The source states are all the points on L different from P, the encoding rules are all the points not on L, and the messages are all the lines not passing through P. Prove the code constructed is a 2-fold perfect Cartesian one of Type I with $|\mathscr{S}| = q, |\mathscr{M}| = |\mathscr{E}| = q^2$ and $P_0 = P_1 = 1/q$ if the encoding rules have a uniform probability distribution. (This code is called modified GMS code.)

3.7 Fix an $(n-1)$-dimensional subspace L of $PG(n, \mathbb{F}_q)$, and fix a point P on L. Source states are all $(n-2)$-dimensional subspaces contained in L, not containing P. Encoding rules are all points not contained in L and the messages are all $(n-1)$-dimensional subspaces which do not contain P. For any source state s and any encoding rule e define $e(s) = <s, e>$. Prove that the resulting code is a 2-fold perfect Cartesian one of Type II with $|\mathscr{S}| = q^{n-1}, |\mathscr{E}| = |\mathscr{M}| = q^n$ and $P_0 = P_1 = 1/q$ if the encoding rules have a uniform probability distribution.

Chapter 4

Authentication Schemes with Arbitration

We have discussed several properties of the authentication schemes with three participants: the information-theoretic bound for the successful probability of spoofing attack, the lower bound for the number of encoding rules, and the characterization of perfect schemes in Chapter 3. In this chapter we will discuss similar problems for the authentication schemes with arbitration.

An authentication code with three participants is called an A-code, while the code with arbitration is called an A^2-code.

4.1 Lower Bounds

By the same method of Proposition 3.1 we can prove the information-theoretic bound for P_{O_r}, P_{R_r}, and P_T, respectively, and find the necessary and sufficient conditions for achieving these bounds.

Define the set

$$\overline{\mathscr{M}_R^r} = \{m^r \in \mathscr{M}^r \mid \mathscr{E}_R(m^r) \neq \emptyset\}.$$

PROPOSITION 4.1

The inequality

$$P_{O_r} \geq 2^{H(E_R|M^{r+1}) - H(E_R|M^r)} \tag{4.1}$$

holds for any integer $r \geq 0$. The equality holds if and only if for any $m^r * m \in \overline{\mathscr{M}_R^{r+1}}$, the ratio

$$\frac{p(f|m^r)}{p(f|m^r * m)} \tag{4.2}$$

is independent of m^r, m, and $f \in \mathscr{E}_R(m^r * m)$. When this equality holds, the probability P_{O_r} equals the constant ratio in Equation (4.2).

PROOF   Let M_{r+1} denote the random variable of the $(r+1)$-th message. For a given $m^r = (m_1, \cdots, m_r) \in \mathcal{M}^r$, let

$$\text{supp}(M_{r+1}, E_R|m^r) = \{(m,f)|p(M_{r+1} = m, f|m^r) > 0, \ m \in \mathcal{M}, f \in \mathscr{E}_R\}$$

denote the support of the conditional probability distribution of the random variable pair (M_{r+1}, E_R) conditional on $M^r = m^r$. Then underbounding a maximum by an average gives

$$\max_{m \in \mathcal{M}} P(m|m^r) \geqslant \sum_{m \in \mathcal{M}} p(M_{r+1} = m|m^r) P(m|m^r) \qquad (4.3)$$

$$= \sum_{(m,f) \in \text{supp}(M_{r+1}, E_R|m^r)} p(M_{r+1} = m|m^r) p(f|m^r)$$

$$= \widetilde{E}\left(\frac{p(M_{r+1} = m|m^r)p(f|m^r)}{p(M_{r+1} = m, f|m^r)}\right)$$

where $\widetilde{E}$ is the expectation on $\text{supp}(M_{r+1}, E_R|m^r)$. By using the Jensen inequality, we obtain

$$\log \max_{m \in \mathcal{M}} P(m|m^r) \geq \widetilde{E}\left(\log \frac{p(M_{r+1} = m|m^r)p(f|m^r)}{p(_{r+1} = m, f|m^r)}\right) \qquad (4.4)$$

$$= H(M_{r+1}, E_R|M^r = m^r) - H(M_{r+1}|M^r = m^r)$$

$$-H(E_R|M^r = m^r)$$

where

$$H(M_{r+1}, E_R|M^r = m^r) = -\sum_{(m,f) \in \text{supp}(M_{r+1}, E_R|m^r)} p(M_{r+1} = m, f|m^r) \log p(M_{r+1} = m, f|m^r),$$

$$H(M_{r+1}|M^r = m^r) = -\sum_{m:p(M_{r+1} = m|m^r) > 0} p(M_{r+1} = m|m^r) \log p(M_{r+1} = m|m^r),$$

$$H(E_R|M^r = m^r) = -\sum_{e \in \mathscr{E}_R(m^r)} p(f|m^r) \log p(f|m^r).$$

Finally, we make another use of the Jensen inequality to obtain

$$\log P_{O_r} = \log \sum_{m^r \in \mathcal{M}^r} p(m^r) \max_{m \in \mathcal{M}} P(m|m^r)$$

$$\geq \sum_{m^r \in \mathcal{M}^r} p(m^r) \log \max_{m \in \mathcal{M}} P(m|m^r) \qquad (4.5)$$

$$\geq H(M_{r+1}, E_R|M^r) - H(M_{r+1}|M^r) - H(E_R|M^r)$$

$$= H(E_R|M^{r+1}) - H(E_R|M^r).$$

Lemma 3.2 is used in the last equality.

From the above derivation, we see that equality holds in this bound if and only if the following two conditions are satisfied:

(i) in order that equalities in Equations (4.3) and (4.4) hold, $P(m|m^r)$ is independent of those m and m^r with $p(M_{r+1} = m \mid m^r) > 0$;

(ii) in order that equalities in Equations (4.4) and (4.5) hold, the ratio

$$\frac{p(M_{r+1} = m|m^r)p(f|m^r)}{p(M_{r+1} = m, f \mid m^r)} = \frac{p(f|m^r)}{p(f|m^r * m)}$$

is independent of m, m^r, and $f \in \mathscr{E}_R(m^r * m)$ when $\mathscr{E}_R(m^r * m) \neq \emptyset$.

Condition (i) can be deduced from condition (ii) since, if $p(M_{r+1} = m \mid m^r) > 0$, then $\mathscr{E}_R(m^r * m) \neq \emptyset$ and

$$P(m|m^r) = \sum_{f \in \mathscr{E}_R(m' * m)} p(f|m^r)$$

$$= \frac{p(f|m^r)}{p(f|m^r * m)} \sum_{f \in \mathscr{E}_R(m' * m)} p(f|m^r * m)$$

$$= \frac{p(f|m^r)}{p(f|m^r * m)}$$

This completes the proof of Proposition 4.1. ⬜

PROPOSITION 4.2

The inequality

$$P_{R_r} \geq 2^{H(E_T|E_R, M^{r+1}) - H(E_T|E_R, M^r)} \tag{4.6}$$

holds for any integer $r \geq 0$. The equality holds if and only if for any $m^r * m \in \mathscr{M}_R^{r+1}$ and $f \in \mathscr{E}_R(m^r * m)$ with $\mathscr{E}_T(f, m^r * m) \neq \emptyset$ the ratio

$$\frac{p(e|f, m^r)}{p(e|f, m^r * m)} \tag{4.7}$$

is independent of m^r, m, $f \in \mathscr{E}_R(m^r * m)$ and $e \in \mathscr{E}_T(f, m^r * m)$. When this equality holds, the probability P_{R_r} equals the constant ratio in Equation (4.7).

PROOF For a given $m^r \in \overline{\mathscr{M}^r}$ and $f \in \mathscr{E}_R(m^r)$, let

$$\text{supp}(M_{r+1}, E_T|f, m^r)$$
$$= \{(m, e)|p(M_{r+1} = m, E_T = e|f, m^r) > 0, \ m \in \mathscr{M}, e \in \mathscr{E}_T\}$$

denote the support of the conditional probability distribution of the random variable pair (M_{r+1}, E_T) conditional on $E_R = f, M^r = m^r$. Then underbounding a maximum by an average gives

$$\max_{m \in \mathcal{M}} P(m|f, m^r) \geqslant \sum_{m \in \mathcal{M}} p(M_{r+1} = m|f, m^r) P(m|f, m^r) \tag{4.8}$$

$$= \sum_{(m,e) \in \text{supp}(M_{r+1}, E_T|f, m^r)} p(M_{r+1} = m|f, m^r) p(e|f, m^r)$$

$$= \widetilde{E}\left(\frac{p(M_{r+1} = m|f, m^r) p(e|f, m^r)}{p(M_{r+1} = m, E_T = e|f, m^r)}\right)$$

where $\widetilde{E}$ is the expectation on $\text{supp}(M_{r+1}, E_T|f, m^r)$. By using the Jensen inequality, we obtain

$$\log \max_{m \in \mathcal{M}} P(m|f, m^r) \geq \log \widetilde{E}\left(\frac{p(M_{r+1} = m|f, m^r) p(e|f, m^r)}{p(M_{r+1} = m, E_T = e|f, m^r)}\right) \tag{4.9}$$

$$= H(M_{r+1}, E_T|E_R = f, M^r = m^r)$$

$$- H(M_{r+1}|E_R = f, M^r = m^r) - H(E_T|E_R = f, M^r = m^r).$$

Furthermore,

$$\log\left(\sum_{m^r} p(m^r|f) \max_{m \in \mathcal{M}} P(m|f, m^r)\right)$$

$$\geq \sum_{m} p(m^r|f) \log \max_{m} P(m|f, m^r) \tag{4.10}$$

$$\geq H(M_{r+1}, E_T|E_R = f, M^r) - H(M_{r+1}|E_R = f, M^r)$$

$$- H(E_T|E_R = f, M^r).$$

Finally we make another use of the Jensen inequality to obtain

$$\log P_{R_r} = \log\left(\sum_{f} p(f) \sum_{m^r} p(m^r|f) \max_{m} P(m|f, m^r)\right)$$

$$\geq \sum_{f} p(f) \log\left(\sum_{m^r} p(m^r|f) \max_{m} P(m|f, m^r)\right) \tag{4.11}$$

$$\geq H(M_{r+1}, E_T|E_R, M^r) - H(M_{r+1}|E_R, M^r) - H(E_T|E_R, M^r)$$

$$= H(E_T|E_R, M^{r+1}) - H(E_T|E_R, M^r).$$

Lemma 3.2 is used in the last equality.

From the above derivation, we see that equality holds in this bound if and only if the following two conditions are satisfied:

(i) in order that equalities in Equations (4.8) and (4.10) hold, $P(m|f, m^r)$ is independent of those m, m^r, and $f \in \mathscr{E}_R(m^r * m)$ when $p(M_{r+1} = m \mid m^r) > 0$;

(ii) in order that equalities in Equations (4.9) and (4.11) hold, the ratio

$$\frac{p(M_{r+1} = m | f, m^r) p(e | f, m^r)}{p(M_{r+1} = m, e \mid f, m^r)} = \frac{p(e | f, m^r)}{p(e | f, m^r * m)}$$

is independent of m, m^r, $f \in \mathscr{E}_R(m^r * m)$ and $e \in \mathscr{E}_T(f, m^r * m)$ when $\mathscr{E}_T(f, m^r * m) \neq \emptyset$.

Condition (i) can be deduced from condition (ii) since, if $p(M_{r+1} = m | f, m^r) > 0$, then $\mathscr{E}_T(f, m^r * m) \neq \emptyset$ and

$$P(m | f, m^r) = \sum_{e \in \mathscr{E}_T(f, m' * m)} p(e | f, m^r)$$

$$= \frac{p(e | f, m^r)}{p(e | f, m^r * m)} \sum_{e \in \mathscr{E}_T(f, m' * m)} p(e | f, m^r * m)$$

$$= \frac{p(e | f, m^r)}{p(e | f, m^r * m)}$$

This completes the proof of Proposition 4.2. ⬜

PROPOSITION 4.3

The inequality

$$P_T \geq 2^{H(E_R | E_T, M') - H(E_R | E_T)} \tag{4.12}$$

holds. The equality holds if and only if for any $e \in \mathscr{E}_T, m' \in \mathscr{M}'(e)$ with $\mathscr{E}_R(e, m') \neq \emptyset$ the ratio

$$\frac{p(f | e)}{p(f | e, m')} \tag{4.13}$$

is independent of e, m' and $f \in \mathscr{E}_R(e, m')$. When this equality holds, the probability P_T equals the constant ratio in Equation (4.13).

PROOF For a given $e \in \mathscr{E}_T$, let

$$\text{supp}(M', E_R | e) = \{(m', f) | m' \in \mathscr{M}'(e), f \in \mathscr{E}_R(e, m'), p(m', f | e) > 0\}$$

denote the support of the conditional probability distribution of the random variable pair (M', E_R) conditional on $E_T = e$. Then underbounding a maximum by an average gives

$$\max_{m'} P(m' | e) \geq \sum_{m'} p(M' = m' | e) P(m' | e) \tag{4.14}$$

$$= \sum_{(m', f) \in \text{supp}(M', E_R | e)} p(M' = m' | e) p(f | e)$$

$$= \tilde{E} \left(\frac{p(M' = m' | e) p(f | e)}{p(M' = m', f | e)} \right)$$

where $\widetilde{E}$ is the expectation on $\text{supp}(M', E_R|e)$. By using the Jensen inequality, we obtain

$$\log \max_{m'} P(m'|e) \geq \log \widetilde{E}\left(\frac{p(M'=m'|e)p(f|e)}{p(M'=m', f|e)}\right) \tag{4.15}$$

$$\geq H(M', E_R|E_T = e) - H(M'|E_T = e) - H(E_T|E_T = e).$$

Finally, we make another use of the Jensen inequality to obtain

$$\log P_T = \log\left(\sum_e p(e) \max_{m'} P(m'|e)\right)$$

$$\geq \sum_e p(e) \log \max_{m'} P(m'|e) \tag{4.16}$$

$$\geq H(M', E_R|E_T) - H(M'|E_T) - H(E_R|E_T)$$

$$= H(E_R|M', E_T) - H(E_R|E_T).$$

Lemma 3.2 is used in the last equality.

From the above derivation, we see that equality holds in this bound if and only if the following two conditions are satisfied:

(i) in order that equalities in Equations (4.14) and (4.16) hold, $P(m'|e)$ is independent of those e, m' when $p(M' = m' \,|\, e) > 0$;

(ii) in order that equalities in Equations (4.15) and (4.16) hold, the ratio

$$\frac{p(M' = m'|e)p(f|e)}{p(M' = m', f \,|\, e)} = \frac{p(f|e)}{p(f|m')}$$

is independent of e, m', and $f \in \mathscr{E}_R(e, m')$ when $\mathscr{E}_R(e, m') \neq \emptyset$.

Condition (i) can be deduced from condition (ii) since, if $p(M' = m' \,|\, e) > 0$, then $\mathscr{E}_R(e, m') \neq \emptyset$ and

$$P(m'|e) = \sum_{f \in \mathscr{E}_R(e,m')} p(f|e)$$

$$= \frac{p(f|e)}{p(f|e, m')} \sum_{f \in \mathscr{E}_R(e,m')} p(f|m')$$

$$= \frac{p(f|e)}{p(f|e, m')}.$$

This completes the proof of Proposition 4.3.   ☐

The following two propositions give the lower bounds of $|\mathscr{E}_R|$ and $|\mathscr{E}_T|$, respectively.

PROPOSITION 4.4

The number of decoding rules of the receiver has the lower bound

$$|\mathscr{E}_R| \geq (P_{O_0} P_{O_1} \cdots P_{O_{t-1}} P_T)^{-1} \tag{4.17}$$

($t \leq k$ is any positive integer.) The number $|\mathscr{E}_R|$ achieves its lower bound in Equation (4.17) if and only if P_T and P_{O_r} ($0 \leq r \leq t-1$) achieve their lower bounds in Equations (4.12) and (4.1),

$$H(E_R|E_T, M') = 0, \quad H(E_R|M^t) = H(E_R|E_T)$$

and E_R has a uniform probability distribution.

PROOF   According to Propositions 4.1 and 4.3, we have

$$P_{O_0} P_{O_1} \cdots P_{O_{t-1}} P_T \geq 2^{H(E_R|M') - H(E_R) + H(E_R|E_T, M') - H(E_R|E_T)}. \tag{4.18}$$

Since

$$H(E_R|E_T, M') \geqslant 0, \tag{4.19}$$

$$H(E_R|M^t) \geqslant H(E_R|E_T) \tag{4.20}$$

(giving E_T is equivalent to giving M^k, and $t \leq k$) and

$$|\mathscr{E}_R| \geqslant 2^{H(E_R)}, \tag{4.21}$$

it follows that

$$|\mathscr{E}_R| \geqslant 2^{H(E_R)} \geqslant (P_{O_0} P_{O_1} \cdots P_{O_{t-1}} P_T)^{-1}.$$

Furthermore the equality in Equation (4.17) holds if and only if the equalities in Equations (4.18), (4.19), (4.20), and (4.21) hold. The equality of Equation (4.21) means that E_R has a uniform probability distribution.   ▯

PROPOSITION 4.5

The number of encoding rules of the transmitter has the lower bound

$$|\mathscr{E}_T| \geq (P_{O_0} P_{O_1} \cdots P_{O_{t-1}} P_{R_0} \cdots P_{R_{t-1}})^{-1}. \tag{4.22}$$

The number $|\mathscr{E}_T|$ achieves its lower bound in Equation (4.22) if and only if P_{O_r}, P_{R_r} ($0 \leq r \leq t-1$) achieve their lower bounds in Equations (4.1) and (4.6),

$$H(E_T|E_R, M^t) = 0, \quad H(E_R|M^t) = H(E_R|E_T)$$

and E_T has a uniform distribution.

PROOF   According to Propositions 4.1 and 4.2, we have

$$P_{O_0} P_{O_1} \cdots P_{O_{t-1}} P_{R_0} P_{R_1} \cdots P_{R_{t-1}}$$

$$\geqslant 2^{H(E_R|M^t)-H(E_R)+H(E_T|E_R,M^t)-H(E_T|E_R)} \tag{4.23}$$

$$= 2^{H(E_R|M^t)-H(E_T,E_R)+H(E_T|E_R,M^t)}$$

$$= 2^{H(E_R|M^t)-H(E_R|E_T)+H(E_T|E_R,M^t)-H(E_T)}$$

By Equation (4.20) and

$$H(E_T|E_R, M^t) \geqslant 0, \tag{4.24}$$

$$|\mathscr{E}_T| \geqslant 2^{H(E_T)}, \tag{4.25}$$

it follows that

$$|\mathscr{E}_T| \geqslant 2^{H(E_T)} \geqslant (P_{O_0} P_{O_1} \cdots P_{O_{t-1}} P_{R_0} P_{R_1} \cdots P_{R_{t-1}})^{-1}. \tag{4.26}$$

Furthermore, the equality in Equation (4.22) holds if and only if the equalities in Equations (4.20), (4.23), (4.24), and (4.25) hold. The equality in Equation (4.25) means that $|E_T|$ has a uniform distribution.   ▯

REMARK 4.1   Since

$$H(E_T|E_R, M^t) = H(E_R, E_T|M^t) - H(E_R|M^t)$$

$$= H(E_T|M^t) + H(E_R|E_T) - H(E_R|M^t),$$

the conditions that $H(E_T|E_R, M^t) = 0$ and $H(E_R|M^t) = H(E_R|E_T)$ are equivalent to that of $H(E_T|M^t) = 0$.   ▯

4.2   Perfect Schemes with Arbitration

Denote an authentication scheme with arbitration by $(\mathscr{S}, \mathscr{M}, \mathscr{E}_R, \mathscr{E}_T, p_S, p_{E_R}, p_{E_T})$.

DEFINITION 4.1   *If an authentication scheme $(\mathscr{S}, \mathscr{M}, \mathscr{E}_R, \mathscr{E}_T, p_S, p_{E_R}, p_{E_T})$ satisfies the following conditions:*

$$|\mathscr{E}_R| = (P_{O_0} P_{O_1} \cdots P_{O_{t-1}} P_T)^{-1}$$

and

$$|\mathscr{E}_T| = (P_{O_0} P_{O_1} \cdots P_{O_{t-1}} P_{R_0} \cdots P_{R_{t-1}})^{-1},$$

then the scheme is called t-fold perfect.

We consider the combinatorial structure of perfect schemes with arbitration. The following Corollaries 4.1, 4.2, and 4.3 will provide a bridge between the information-theoretic lower bounds of P_{O_r}, P_{R_r}, and P_T and the combinatorial structure of perfect schemes.

COROLLARY 4.1

Suppose that E_R has a uniform probability distribution and

$$P_{O_r} = 2^{H(E_R|M^{r+1}) - H(E_R|M^r)}, \ 0 \le r \le t - 1 \qquad (4.27)$$

then for any $m^r * m \in \overline{\mathscr{M}_R^{r+1}}$, we have

$$P_{O_r} = |\mathscr{E}_R(m^r * m)| / |\mathscr{E}_R(m^r)|, \ \ 0 \le r \le t - 1.$$

Here, $\mathscr{E}_R(m^0) = \mathscr{E}_R$.

PROOF If $\mathscr{E}_R(m^r * m) \ne \emptyset$ it follows from Proposition 4.1 and the assumption in Equation (2.1) that for any $f \in \mathscr{E}_R(m^r * m)$,

$$
\begin{aligned}
P_{O_r} &= \frac{p(f|m^r)}{p(f|m^r * m)} \\
&= \frac{p(f, m^r)p(m^r * m)}{p(f, m^r * m)p(m^r)} \\
&= p(f) \sum_{e \in \mathscr{E}_T(f, m^r)} p(e|f)p(f(m^r)) \\
&\quad \times \left(p(f) \sum_{e \in \mathscr{E}_T(f, m^r * m)} p(e|f)p(f(m^r * m)) \right)^{-1} \\
&\quad \times \sum_{f' \in \mathscr{E}_R(m^r * m)} p(f') \sum_{e' \in \mathscr{E}_T(f', m^r * m)} p(e'|f')p(f'(m^r * m)) \\
&\quad \times \left(\sum_{f' \in \mathscr{E}_R(m^r)} p(f') \sum_{e' \in \mathscr{E}_T(f', m^r)} p(e'|f')p(f'(m^r)) \right)^{-1}.
\end{aligned}
\qquad (4.28)
$$

Now we prove by induction on r that the summation

$$\sum_{e \in \mathscr{E}_T(f, m^r * m)} p(e|f)p(f(m^r * m)), \ \ 0 \le r \le t - 1 \qquad (4.29)$$

does not depend on $f \in \mathscr{E}_R(m^r * m)$.

When $r = 0$, we have from Equation (4.28) that

$$P_{O_0} = \frac{p(f)p(m)}{p(f,m)}$$

$$= \frac{\sum\limits_{f' \in \mathscr{E}_R(m)} p(f') \sum\limits_{e' \in \mathscr{E}_T(f',m)} p(e'|f')p(f'(m))}{\sum\limits_{e \in \mathscr{E}_T(f,m)} p(e|f)p(f(m))}$$

which does not depend on $f \in \mathscr{E}_R(m)$, therefore, the summation in its denominator also does not depend on f. Now we suppose that the summation of Equation (4.29) does not depend on $f \in \mathscr{E}_R(m^u)$ for $u < t - 1$, then it follows also from Equation (4.28) that

$$P_{O_u} = \sum\limits_{f' \in \mathscr{E}_R(m^u * m)} p(f') \sum\limits_{e \in \mathscr{E}_T(f', m^u * m)} p(e|f')p(f'(m^u * m))$$

$$\times \left(\sum\limits_{e \in \mathscr{E}_T(f, m^u * m)} p(e|f)p\big(f(m^u * m)\big) \sum\limits_{f' \in \mathscr{E}_R(m^u)} p(f') \right)^{-1}$$

which does not depend on $f \in \mathscr{E}_R(m^u * m)$. Hence, the summation of Equation (4.29) for $r = u + 1$ does not depend on $f \in \mathscr{E}_R(m^u * m)$.

It follows from Equation (4.28) that

$$P_{O_r} = \frac{\sum\limits_{f \in \mathscr{E}_R(m^r * m)} p(f)}{\sum\limits_{f \in \mathscr{E}_R(m^r)} p(f)} = \frac{|\mathscr{E}_R(m^r * m)|}{|\mathscr{E}_R(m^r)|}$$

where the second equality uses the assumption that E_R has a uniform distribution. ▯

For a given $f \in \mathscr{E}_R$, let $E_T(f)$ denote the random variable of valid encoding rules of f; it takes its values from $\mathscr{E}_T(f)$. Similarly, for a given $e \in \mathscr{E}_T$, let $E_R(e)$ denote the random variable of decoding rules of which e is valid; it takes its values from $\mathscr{E}_R(e)$.

COROLLARY 4.2

For any given $f \in \mathscr{E}_R$, suppose that $E_T(f)$ has a uniform probability distribution and

$$P_{R_r} = 2^{H(E_T|E_R, M^{r+1}) - H(E_T|E_R, M^r)}, 0 \leq r \leq t - 1. \qquad (4.30)$$

Then for any $m^r * m \in \overline{\mathscr{M}_R^{r+1}}$, $f \in \mathscr{E}_R(m^r * m)$,

$$P_{R_r} = \frac{|\mathscr{E}_T(f, m^r * m)|}{|\mathscr{E}_T(f, m^r)|}.$$

PROOF   If $\mathscr{E}_T(f, m^r \star m) \neq \emptyset$, it follows from Proposition 4.2 and the assumption in Equation (2.1) that for any $e \in \mathscr{E}_T(f, m^r \star m)$,

$$
\begin{aligned}
P_{R_r} &= \frac{p(e|f, m^r)}{p(e|f, m^r * m)} \\
&= \frac{p(e, f, m^r)p(f, m^r * m)}{p(e, f, m^r * m)p(f, m^r)} \\
&= p\big(f(m^r)\big) \sum_{e' \in \mathscr{E}_T(f, m^r * m)} p(e'|f)p\big(f(m^r * m)\big) \\
&\quad \times \left(p\big(f(m^r * m)\big) \sum_{e' \in \mathscr{E}_T(f, m^r)} p(e'|f)p\big(f(m^r)\big) \right)^{-1} \\
&= \frac{\sum_{e' \in \mathscr{E}_T(f, m^r * m)} p(e'|f)}{\sum_{e' \in \mathscr{E}_T(f, m^r)} p(e'|f)} \\
&= \frac{|\mathscr{E}_T(f, m^r * m)|}{|\mathscr{E}_T(f, m^r)|}.
\end{aligned}
$$

The last equality is due to the assumption that $\mathscr{E}_T(f)$ has a uniform probability distribution.   ▯

COROLLARY 4.3

For any given $e \in \mathscr{E}_T$, suppose that $E_R(e)$ has a uniform probability distribution and

$$
P_T = 2^{H(E_R|E_T, M') - H(E_R|E_T)}.
$$

Then for any $e \in \mathscr{E}_T$, $m' \in \mathscr{M}'(e)$ with $\mathscr{E}_R(e, m') \neq \emptyset$,

$$
P_T = \frac{|\mathscr{E}_R(e, m')|}{|\mathscr{E}_R(e)|}.
$$

PROOF   It follows from the proof of Proposition 4.3 that for any $e \in \mathscr{E}_T$, $m' \in \mathscr{M}'(e)$ with $\mathscr{E}_R(e, m') \neq \emptyset$,

$$
P_T = P(m'|e) = \sum_{f \in \mathscr{E}_R(e, m')} p(f|e) = |\mathscr{E}_R(e, m')|/|\mathscr{E}_R(e)|.
$$

▯

The perfect schemes have a requirement for the probability distributions of $\mathscr{S}^r$.

COROLLARY 4.4

Suppose the equalities in Equations (4.27) and (4.30) hold. Then for any $m^r \in \overline{\mathscr{M}_R^r}$, the probability $p(f[m^r])$ does not depend on $f \in \mathscr{E}_R(m^r)$ ($1 \leqslant r \leqslant t$).

PROOF   In the proof of Corollaries 4.1 and 4.2 it is shown that

$$\sum_{e \in \mathscr{E}_T(f, m^r)} p(e|f)p\big(f(m^r)\big)$$

does not depend on $f \in \mathscr{E}_R(m^r)$ and

$$\sum_{e \in \mathscr{E}_T(f, m^r)} p(e|f) = P_{R_0} P_{R_1} \cdots P_{R_{r-1}} \times \sum_{e \in \mathscr{E}_T(f)} p(e|f)$$

$$= P_{R_0} P_{R_1} \cdots P_{R_{r-1}}.$$

The conclusion follows immediately.   ▯

Now we show that for a perfect scheme the number $|\mathscr{M}(f, s)|$ is a constant for any $f \in \mathscr{E}_R$ and $s \in S$.

COROLLARY 4.5

If $P_{R_0} = 2^{H(E_T|E_R, M) - H(E_T|E_R)}$, then $|\mathscr{M}(f, s)| = c$ is a constant for any $f \in \mathscr{E}_R$ and $s \in \mathscr{S}$. Furthermore, $c = P_{R_0}^{-1}$.

PROOF   Given a decoding rule f, we have assumed that for any $m \in \mathscr{M}(f)$ there exists at least one encoding rule e such that $m \in \mathscr{M}(e)$. Hence $P(m|f) > 0$ for any $m \in \mathscr{M}(f)$. If P_{R_0} achieves its information-theoretic lower bound, then P_{R_0} is equal to $P(m|f)$ for any $f \in \mathscr{E}_R$ and $m \in \mathscr{M}(f)$ from the proof of Proposition 4.2. Since

$$\sum_{m \in \mathscr{M}(f,s)} P(m|f) = \sum_{m \in \mathscr{M}(f,s)} \sum_{e \in \mathscr{E}_T(f,m)} p(e|f)$$

$$= \sum_{e \in \mathscr{E}_t(f)} p(e|f) \sum_{m:e(m)=s} 1$$

$$= \sum_{e \in \mathscr{E}_T(f)} p(r|f) = 1,$$

for any $s \in \mathscr{S}$, it follows that $|\mathscr{M}(f,s)| = P(m|f)^{-1} = P_{R_0}^{-1}$.   ▯

In order to study the combinatorial characterization of perfect authentication schemes with arbitration, we introduce the definition of restricted partially balanced designs (**RPBD**).

DEFINITION 4.2   *Let v, b, k, c, λ, t be positive integers. A PBD t-design $(\mathcal{M}, \mathcal{E})$ is called restricted (RPBD t-design) if $|\mathcal{M}| = v, \mathcal{E} = (\mathcal{E}_1, \cdots, \mathcal{E}_b)$, each block $\mathcal{E}_i$ is divided into k groups, and each group has c points of $\mathcal{M}$. Any t-subset of $\mathcal{M}$ either occurs in exactly λ blocks in such a way that each point of the t-subset occupies one group, or does not exist in such blocks at all.*

Similarly, we can introduce restricted strong partially balanced design (**RS PBD**) $t - (v, b, k, c; \lambda_1, \lambda_2, \cdots, \lambda_t, 0)$.

Remember that for a given $f \in \mathcal{E}_R$ and $e \in \mathcal{E}_T(f)$,

$$\mathcal{M}'(e) = \mathcal{M} \setminus \mathcal{M}(e)$$

and

$$\mathcal{M}'_f(e) = \mathcal{M}(f) \setminus \mathcal{M}(e) \subset \mathcal{M}'(e).$$

Our main result about the perfect authentication schemes with arbitration is the following theorem.

THEOREM 4.1

The necessary and sufficient conditions for a scheme $\mathcal{A} = (\mathcal{S}, \mathcal{M}, \mathcal{E}_R, \mathcal{E}_T, p_S, p_{E_R}, p_{E_T})$ being t-fold perfect are as follows:

(i) *$E_T, E_R, E_T(f)(\forall f \in \mathcal{E}_R)$ and $E_R(e)(\forall e \in \mathcal{E}_T)$, have uniform probability distribution.*

(ii) *For any given $m^r \in \overline{\mathcal{M}_R^r}(1 \leq r \leq t)$, the probability $p(S^r = f[m^r])$ is constant for all $f \in \mathcal{E}_R(m^r)$.*

(iii) *For any given $e \in \mathcal{E}_T$ the pair*

$$\{\mathcal{M}'(e), [\mathcal{M}'_f(e)|f \in \mathcal{E}_R(e)]\}$$

is a $1 - (v - k, b_1, k[c - 1]; 1, 0)$ design where $v = |\mathcal{M}|$, $k = |\mathcal{S}|$, $b_1 = |\mathcal{E}_R(e)|$, c is a positive integer (in fact, $c = |\mathcal{M}(f, s)|$ for all $f \in \mathcal{E}_R, s \in \mathcal{S}$).

(iv) *For any given $f \in \mathcal{E}_R$, the pair*

$$\{\mathcal{M}(f), [\mathcal{M}(e)|e \in \mathcal{E}_T(f)]\}$$

is a SPBD $t - (kc, b_2, k; \lambda_1, \cdots, \lambda_t, 0)$ where $\lambda_t = 1$, $\lambda_r = (P_{R_r} \cdots P_{R_{t-1}})^{-1}, 1 \leq r \leq t - 1, b_2 = |\mathcal{E}_T(f)|$.

(v) *The pair*

$$\{\mathcal{M}, [\mathcal{M}(f)|f \in \mathcal{E}_R]\}$$

is a RSPBD $t - (v, b_3, k, c; \mu_1, \cdots, \mu_t, 0)$, where

$$\mu_t = P_T^{-1}, \mu_r = (P_{O_r} \cdots P_{O_{t-1}} P_T)^{-1}, 1 \leq r \leq t - 1, b_3 = |\mathcal{E}_R|.$$

PROOF   Suppose that the scheme $\mathscr{A}$ is t-fold perfect. We prove that the conditions (i) through (v) are satisfied.

According to Propositions 4.4 and 4.5, it follows that P_{O_r}, P_{R_r} for $0 \le r \le t-1$ and P_T achieve their lower bounds in Equations (4.1), (4.6), and (4.12), respectively, and E_T and E_R have uniform probability distribution. It will be proved that $E_T(f)$ and $E_R(e)$ also have uniform probability distribution later. The condition (ii) is nothing but Corollary 4.4.

For any given $e \in \mathscr{E}_T$ we have $|\mathscr{M}'(e)| = v - k$ and $|\mathscr{M}'_f(e)| = k(c-1)$ for any $f \in \mathscr{E}_R(e)$ from Corollary 4.5. Since

$$H(E_R | E_T, M') = - \sum_{f \in \mathscr{E}_R(e,m')} p(f,e,m') \log p(f|e,m') = 0$$

from Proposition 4.4, it is deduced that $\mathscr{E}_R(e,m') = \emptyset$ or $|\mathscr{E}_R(e.m')| = 1$ for any m', hence (iii) is followed.

Therefore, for any $e \in \mathscr{E}_T$, and $m' \in \mathscr{M}'(e)$ with $\mathscr{E}_R(e,m') \ne \emptyset$, there exists a unique $f \in \mathscr{E}_R(e,m')$. Hence, we have $p(f|e,m') = 1$ and $P_T = p(f|e)$ by Proposition 4.3. Thus, $E_R(e)$ has also a uniform probability distribution. Since

$$p(e|f) = \frac{p(e)p(f|e)}{p(f)},$$

$E_T(f)$ has also a uniform probability distribution. Thus, the condition (i) is satisfied.

For any given $f \in \mathscr{E}_R$ and $e \in \mathscr{E}_T(f)$, we have $|\mathscr{M}(f)| = kc$ and $|\mathscr{M}(e)| = k$. Since $H(E_T | E_R, M^t) = 0$ from Proposition 4.5, it is deduced that $|\mathscr{E}_T(f,m^t)| = 0$ or 1 by the same argument used above. This means that $\lambda_t = 1$. Suppose that $|\mathscr{E}_T(f,m^t)| = 1$ for some m^t, it follows that

$$|\mathscr{E}_T(f)| = \left(P_{R_0} \cdots P_{R_{t-1}} \right)^{-1} \qquad (4.31)$$

from Corollary 4.2 and the fact that $E_T(f)$ has a uniform probability distribution. Assume that $\mathscr{E}_T(f,m^r) \ne \emptyset$ for some $m^r, 1 \le r \le t-1$. Taking any $e \in \mathscr{E}_T(f,m^r)$ we can find $m^r \subset m^t \subset \mathscr{M}(e)$, it follows also that $|\mathscr{E}_T(f,m^t)| = 1$ and

$$\lambda_r = |\mathscr{E}_T(f,m^r)| = \left(P_{R_r} \cdots P_{R_{t-1}} \right)^{-1}$$

from Corollary 4.2. This proves (iv).

Take any $m^r = (m_1, \cdots, m_r)$, $r \le t$, then $m_i \in \mathscr{M}(f)$, $1 \le i \le r$, and each m_i occupies one different $\mathscr{M}(f,s)$ if and only if $f \in \mathscr{E}_R(m^r)$. Since E_R has a uniform probability distribution, it follows from Corollary 4.1 that

$$|\mathscr{E}_R(m^t)| = |\mathscr{E}_R| P_{O_0} \cdots P_{O_{t-1}} = P_T^{-1}$$

when $\mathscr{E}_R(m^t) \ne \emptyset$. This means that $\mu_t = P_T^{-1}$. If $\mathscr{E}_R(m^r) \ne \emptyset$ for some $m^r, 1 \le r \le t-1$, it is deduced also that

$$\mu_r = |\mathscr{E}_R(m^r)| = (P_{O_r} \cdots P_{O_{t-1}} P_T)^{-1}$$

from Corollary 4.1. This proves (v).

Conversely, we suppose that the conditions (i) through (v) hold and prove that the scheme $\mathscr{A}$ is perfect. {Here we do not need to assume that λ_r and μ_r for $1 \leq r \leq t-1$ have their expressions in (iv) and (v).} Since the transmitter knows one encoding rule in $\mathscr{E}_T(f)$, which means that he knows t messages in $\mathscr{M}(f)$, therefore, $P_T = \mu_t^{-1}$.

Take any $m^r * m \in \mathscr{M}_R^{r+1}$ and $f \in \mathscr{E}_R(m^r * m), 0 \leq r \leq t-1$. Using the conditions (i), (ii), and (v) and the proof of Corollary 4.1, it follows that

$$P_{O_0} = \frac{p(f)}{p(f|m)} = \frac{|\mathscr{E}_R(m)|}{|\mathscr{E}_R|} = \frac{\mu_1}{|\mathscr{E}_R|}$$

and

$$P_{O_r} = \frac{p(f|m^r)}{p(f|m^r * m)} = \frac{|\mathscr{E}_R(m^r * m)|}{|\mathscr{E}_R(m^r)|} = \frac{\mu_{r+1}}{\mu_r}, \quad 1 \leq r \leq t-1.$$

Hence

$$|\mathscr{E}_R| = (P_{O_0} \cdots P_{O_{t-1}})^{-1} \mu_t = (P_{O_0} \cdots P_{O_{t-1}} P_T)^{-1}.$$

This means that $|\mathscr{E}_R|$ achieves its lower bound.

Similarly, by (iv) we calculate

$$P_{R_0} = \frac{p(e|f)}{p(e|f,m)} = \frac{|\mathscr{E}_T(f,m)|}{b_2} = \frac{\lambda_1}{|\mathscr{E}_T(f)|},$$

$$P_{R_r} = \frac{p(e|f,m^r)}{p(e|f,m^r * m)} = \frac{|\mathscr{E}_T(f,m^r * m)|}{|\mathscr{E}_T(f,m^r)|} = \frac{\lambda_{r+1}}{\lambda_r}, \quad 1 \leq r \leq t-1$$

for any $f \in \mathscr{E}_R, m^r * m \in \mathscr{M}(f)$ and $e \in \mathscr{E}_T(f, m^r * m)$. Therefore,

$$|\mathscr{E}_T(f)| = \left(P_{R_0} \cdots P_{R_{t-1}}\right)^{-1}.$$

By (iii) we have

$$P_T = \sum_{f \in \mathscr{E}_R(e,m')} p(f|e) = \frac{1}{|\mathscr{E}_R(e)|} \tag{4.32}$$

for any $e \in \mathscr{E}_T$, $m' \in \mathscr{M}'(e)$.

Hence,

$$|\mathscr{E}_T| = \frac{|\mathscr{E}_R||\mathscr{E}_T(f)|}{|\mathscr{E}_R(e)|} = (P_{O_0} \cdots P_{O_{t-1}} P_{R_0} \cdots P_{R_{t-1}})^{-1},$$

which means that $|\mathscr{E}_T|$ achieves its lower bound. This completes the proof of Theorem 4.1. ☐

4.3   Perfect Cartesian A^2-Codes

Similar to Cartesian A-code, an A^2-code is Cartesian if one can always know the source state from the message sent by the transmitter. In a Cartesian A^2-code for any $m \in \mathscr{M}$ there is a unique $s \in \mathscr{S}$ such that $m \in \mathscr{M}(f,s)$ for all $f \in \mathscr{E}_R(m)$. Let

$$\mathscr{M}(s) = \{m \in \mathscr{M} | m \in \mathscr{M}(f,s) \ for \ some \ f \in \mathscr{E}_R\}.$$

It is clear that

$$\{\mathscr{M}(s) | s \in \mathscr{S}\}$$

are mutually disjointed and

$$\mathscr{M} = \bigcup_{s \in \mathscr{S}} \mathscr{M}(s).$$

COROLLARY 4.6
In a perfect Cartesian A^2-code, we have $|\mathscr{M}(s)| = v/k$ for all $s \in \mathscr{S}$.

PROOF   For any given $s \in \mathscr{S}$ we have

$$\sum_{m \in \mathscr{M}(s)} \sum_{f \in \mathscr{E}_R(m)} p(f) = \sum_{f \in \mathscr{E}_R} p(f) \sum_{m \in \mathscr{M}(f,s)} 1 = c \cdot \sum_{f \in \mathscr{E}_R} p(f)$$

and

$$P_{O_0} = \frac{\displaystyle\sum_{f \in \mathscr{E}_R(m)} p(f)}{\displaystyle\sum_{f \in \mathscr{E}_R} p(f)}$$

from Corollary 4.1. Hence,

$$|\mathscr{M}(s)| \cdot P_{O_0} = c,$$

$|\mathscr{M}(s)|$ is a constant for all $s \in \mathscr{S}$. Furthermore, we have

$$|\mathscr{M}(s)| \times k = v.$$

This proves the corollary.   □

In a perfect Cartesian A^2-code, there are $l = v/k$ messages in $\mathscr{M}(s)$ for each source state s and there are $c = |\mathscr{M}(f,s)|$ valid messages in $\mathscr{M}(s)$ for each decoding rule f and each source state s. Denote the set of source states by $\mathscr{S} = \{s_1, s_2, \cdots, s_k\}$, and let $s_{i_1}, s_{i_2}, \cdots, s_{i_r}$ be any r-subset of $\mathscr{S}$, put

$$\mathscr{M}_{i_1, i_2, \cdots, i_r} = \{m^r = (m_1, m_2, \cdots, m_r) | m_h \in \mathscr{M}(s_{i_h}), 1 \le h \le r\}$$

and

$$\mathscr{M}_{f,i_1,\cdots,i_r} = \{m_f^r = \{m_1, m_2, \cdots, m_r\} | m_h \in \mathscr{M}(f, s_{i_h}), 1 \le h \le r\}$$

for $f \in \mathscr{E}_R$. We have

$$|\mathscr{M}_{i_1,i_2,\cdots,i_r}| = l^r, \quad |\mathscr{M}_{f,s_1,s_2,\cdots,s_r}| = c^r.$$

Define

$$\mathscr{E}_T(m^r) = \{e \in \mathscr{E}_T | m^r \subset \mathscr{M}(e)\}.$$

COROLLARY 4.7

In a perfect Cartesian A^2-code, assume that for any t-subset $s_{i_1}, s_{i_2}, \cdots, s_{i_t}$ of $\mathscr{S}$ and any $m^t \in \mathscr{M}_{i_1,i_2,\cdots,i_t}$ we have $|\mathscr{E}_T(m^t)| = 1$. Then

(i) *$|\mathscr{E}_T| = l^t$;*

(ii) *for any given $f \in \mathscr{E}_R$, $|\mathscr{E}_T(f, m_f^t)| = 1$ for all $m_f^t \in \mathscr{M}_{f,i_1,i_2,\cdots,i_t}$;*

(iii) *$|\mathscr{E}_R(m^t)| = \mu_t = P_T^{-1}$ for all $m^t \in \mathscr{M}_{i_1,i_2,\cdots,i_t}$;*

(iv) *$P_{R_0} = P_{R_1} = \cdots = P_{R_{t-1}} = 1/c$;*

(v) *$P_{O_0} = P_{O_1} = \cdots = P_{O_{t-1}} = c/l$;*

(vi) *$P_T = (c-1)/(l-1)$;*

(vii) *$|\mathscr{E}_R| = l^t(l-1)/c^t(c-1)$, $|\mathscr{E}_T(f)| = c^t$ for any $f \in \mathscr{E}_R$, and $|\mathscr{E}_R(e)| = (l-1)/(c-1)$ for any $e \in \mathscr{E}_T$;*

(viii) *for any given $e \in \mathscr{E}_T$ and $m \in \mathscr{M}'(e)$, there exists a unique $f \in \mathscr{E}_R(e)$ such that $m \in \mathscr{M}_f'(e)$, i.e., $|\mathscr{E}_R(e, m)| = 1|$.*

PROOF   Each encoding rule e corresponds to a t-subset $m^t = \{e(s_1), e(s_2), \cdots, e(s_t)\} \in \mathscr{M}_{i_1,i_2,\cdots,i_t}$, hence (i) is proved by the assumption of the corollary. For any given $f \in \mathscr{E}_R$ and any $m_f^t \in \mathscr{M}_{f,i_1,i_2,\cdots,i_t}$, since $|\mathscr{E}_T(m_f^t)| = 1$ by the assumption, (ii) follows immediately. For any $m^t \in \mathscr{M}_{i_1,i_2,\cdots,i_t}$, the set $\mathscr{E}_R(m^t)$ is not empty since there exists an encoding rule $e \in \mathscr{E}_T(m^t)$; thus (iii) is held by (v) of Theorem 4.1.

Given $f \in \mathscr{E}_R$ and $m^r = \{m_1, m_2, \cdots, m_r\} \in \mathscr{M}_{f,i_1,i_2,\cdots,i_r}, 0 \le r \le t$, there exist c^{t-r} choices of $\{m_{r+1}, \cdots, m_t\}$ such that $\{m^r, m_{r+1}, \cdots, m_t\} \in \mathscr{M}_{f,i_1,i_2,\cdots,i_t}$, hence $|\mathscr{E}_T(f, m^r)| = c^{t-r}$ by (ii). This proves (iv) by Corollary 4.2.

For $m^r \in \mathscr{M}_{f,i_1,\cdots,i_r}, 0 \le r \le t$, we have $\mathscr{E}_R(m^r) \ne \emptyset$ by (iii). Suppose $s \ne s_h, 1 \le h \le r$, we have

$$\sum_{m \in \mathscr{M}(s)} \sum_{f \in \mathscr{E}_R(m^r * m)} p(f) = \sum_{f \in \mathscr{E}_R(m^r)} p(f) \sum_{m \in \mathscr{M}(f,s)} 1 = c \cdot \sum_{f \in \mathscr{E}_R(m^r)} p(f),$$

and

$$P_{O_r} = \frac{\sum\limits_{f \in \mathscr{E}_R(m' * m)} p(f)}{\sum\limits_{f \in \mathscr{E}_R(m')} p(f)}$$

by Corollary 4.1. Hence,

$$|\mathscr{M}(s)| \cdot P_{O_r} = c.$$

This proves (v).

For any given $e \in \mathscr{E}_T$,

$$\sum_{m' \in \mathscr{M}'(e)} \sum_{f \in \mathscr{E}_R(e,m')} p(f|e) = \sum_{f \in \mathscr{E}_R(e)} p(f|e) \sum_{m' \in \mathscr{M}'_f(e)} 1 = (kc - k) \sum_{f \in \mathscr{E}_R(e)} p(f|e)$$

and

$$P_T = \frac{\sum\limits_{f \in \mathscr{E}_R(e,m')} p(f|e)}{\sum\limits_{f \in \mathscr{E}_R(e)} p(f|e)}$$

by Corollary 4.3, hence,

$$(v - k)P_T = k(c - 1).$$

This proves (vi). Condition (vii) follows from (iv) through (vi) and equalities in Equations (4.31) and (4.32). Since

$$|\mathscr{M}'_f(e)| \cdot |\mathscr{E}_R(e)| = (ck - k) \cdot \frac{l-1}{c-1} = v - k = |\mathscr{M}'(e)|,$$

(viii) follows from (iii) of Theorem 4.1.   ☐

We call the perfect Cartesian A^2-codes, which satisfy the assumption of Corollary 4.7, of Type I, otherwise it is called of Type II. In the following we will see an example of Type II, for which the property (v) of Corollary 4.7 does not hold.

Now we look at an example of the 2-fold perfect Cartesian A^2-code of Type I. We check that it satisfies the condition given in Theorem 4.1.

We will use the projective space $PG(3, \mathbb{F}_q)$. By Lemma 3.5 we know that the number of points in $PG(3, \mathbb{F}_q)$ is

$$N(3, 0) = q^3 + q^2 + q + 1,$$

the number of lines in $PG(3, \mathbb{F}_q)$ is

$$N(3, 1) = q^4 + q^3 + 2q^2 + q + 1,$$

and the number of planes in $PG(3, \mathbb{F}_q)$ is

$$N(3, 2) = q^3 + q^2 + q + 1.$$

From Corollary 3.5, the number of points on a line in $PG(3, \mathbb{F}_q)$ is

$$N(1, 0) = q + 1,$$

and the number of points on a plane in $PG(3, \mathbb{F}_q)$ is

$$N(2, 0) = q^2 + q + 1.$$

From Lemma 3.6, the number of lines passing through a fixed point is

$$Q_3(0, 1) = q^2 + q + 1,$$

the number of planes containing a fixed point is

$$Q_3(0, 2) = q^2 + q + 1,$$

and the number of planes containing a fixed line is

$$Q_3(1, 2) = q + 1.$$

We can see that any two different lines may intersect at one point or not intersect. A line and a plane may intersect at one point or the line is contained in the plane. Any two different planes always intersect at one line.

Scheme A

Fix a line L_0 in $PG(3, \mathbb{F}_q)$. The points on L_0 are regarded as source states. The receiver's decoding rule f is a point not on L_0. The transmitter's encoding rule is a line e not intersecting L_0. An encoding rule e is valid under a decoding rule f if and only if the point f is on the line e. A source state s is encoded by an encoding e into the message $e(s) = \langle e, s \rangle$ which is the unique plane passing through e and s. The receiver accepts a message if and only if the decoding rule f undertaken is contained in the received plane, and finds the corresponding source state by checking the intersect point of the message with the line L_0. Since each message has only one fixed intersect point with the line L_0, this code is Cartesian.

Let us look at the combinatorial structure of this A^2-code. There are $q + 1$ points on L_0, thus $|\mathscr{S}| = q + 1$. The messages are all planes intersecting the line L_0 in one point. This is the same as all planes do not contain the fixed line L_0. The total number of planes is $q^3 + q^2 + q + 1$ and the number of planes containing the line L_0 is $q + 1$. Thus, $|\mathscr{M}| = q^3 + q^2$. The receiver's decoding rules are all points not on L_0. The total number of points is $q^3 + q^2 + q + 1$, thus $|\mathscr{E}_R| = q^3 + q^2$. The transmitter's encoding rules are all lines not intersecting L_0. The total number of lines is $q^4 + q^3 + 2q^2 + q + 1$, among which $(q + 1)(q^2 + q) + 1 = q^3 + 2q^2 + q + 1$ lines intersect L_0 (including L_0 itself.) So $|\mathscr{E}_T| = q^4$. For a given decoding rule f, the encoding rules which are valid under f are all lines passing through the point f but not intersecting

L_0. The total number of lines passing through f is $q^2 + q + 1$, among which $q+1$ lines intersect L_0. Thus, $|\mathscr{E}_T(f)| = q^2$. The decoding rules under which a given encoding rule e is valid are all points on the line e , thus, $|\mathscr{E}_R(e)| = q+1$.

Given a source state s, there are $q^2 + q + 1 - q - 1 = q^2$ planes passing through s but not containing L_0, hence $l = |\mathscr{M}(s)| = q^2$. Given a decoding rule f and a source state s, there are q planes passing through f and s but not containing L_0. Thus, $c = |\mathscr{M}(f,s)| = q$ and $|\mathscr{M}(f)| = q(q+1)$.

Given an encoding rule $e \in \mathscr{E}_T$, any message $m \in \mathscr{M}'(e)$ is a plane which does not contain the line e, hence it intersects e at a unique point which can be taken as a decoding rule f. This means that m is contained in a unique $\mathscr{M}'_f(e)$. Thus, $\{\mathscr{M}'(e), [\mathscr{M}'_f(e)|f \in \mathscr{E}_R(e)]\}$ is a $1-(q^3+q^2-q-1, q+1, q^2-1)$ design.

Given a decoding rule $f \in \mathscr{E}_R$, any message $m \in \mathscr{M}(f)$ is a plane containing f and intersecting L_0 at a unique point. We can find by the similar argument used above that there are $q + 1$ lines passing through f in this plane, among which one line intersecting L_0. Hence, there are q lines passing through f but not intersecting L_0 in the plane. This means that there are q encoding rules $e \in \mathscr{E}_T(f)$ such that $m \in \mathscr{M}(e)$. Any two messages $m_1 \in \mathscr{M}(f,s_1)$ and $m_2 \in \mathscr{M}(f,s_2)$ with $s_1 \neq s_2$ are two planes passing through f and intersecting L_0 at different points s_1 and s_2, respectively. These two planes have a unique common line passing through f and not intersecting L_0. Thus m_1 and m_2 are contained in a unique $\mathscr{M}(e), \{e \in \mathscr{E}_T(f)\}$ (this also means that the scheme satisfies the assumption of Corollary 4.7). Any two messages $m_1, m_2 \in \mathscr{M}(f,s)$ are two planes passing through f and s. The common line between them is the line connecting f and s. Hence, m_1 and m_2 could not be contained in one $\mathscr{M}(e)(e \in \mathscr{E}_T[f])$. Thus, $\{\mathscr{M}(f), [\mathscr{M}(e))|e \in \mathscr{E}_T(f)]\}$ is a SPBD $2\text{-}(q[q + 1], q^2, q + 1; q, 1, 0)$ design.

Any message m is a plane containing $q^2 + q$ points not on L_0. Hence, $m \in \mathscr{M}(f)$ for $q^2 + q$ decoding rules f. For any two messages m_1 and m_2, if they are two planes intersecting L_0 at two different points s_1 and s_2, then they intersect at a line which is not L_0, so there are $q+1$ decoding rules f such that $m_1 \in \mathscr{M}(f,s_1)$ and $m_2 \in \mathscr{M}(f,s_2)$; if they are two planes intersecting L_0 at the same point, then there are no decoding rules f such that $m_1 \in \mathscr{M}(f,s_1)$ and $m_2 \in \mathscr{M}(f,s_2)$ with different s_1 and s_2. Thus, $\{\mathscr{M}, [\mathscr{M}(f)|f \in \mathscr{E}_R]\}$ is a RSPBD $2\text{-}(q^3 + q^2, q^3 + q^2, q + 1, q; q^2 + q, q + 1, 0)$ design.

The conditions (iii), (iv), and (v) of Theorem 4.1 are satisfied by this A^2-code. If we suppose that the condition (i) of Theorem 4.1 is also satisfied, then this A^2-code becomes 2-fold perfect. We can find that

$$P_{O_0} = \frac{q^2 + q}{q^3 + q^2} = \frac{1}{q},$$

$$P_{O_1} = \frac{q + 1}{q^2 + 1} = \frac{1}{q},$$

$$P_{R_0} = \frac{q}{q^2} = \frac{1}{q}, \; P_{R_1} = \frac{1}{q}, P_T = \frac{1}{q + 1},$$

$$|\mathscr{E}_R| = q^3 + q^2 = (P_{O_0}P_{O_1}P_T)^{-1},$$
$$|\mathscr{E}_T| = q^4 = (P_{O_0}P_{O_1}P_{R_0}P_{R_1})^{-1},$$

as given in Corollary 4.7. This is a perfect Cartesian A^2-code of Type I.

The following is an example of 2-fold perfect Cartesian A^2-codes of Type II which is based on the affine space $AG(4, \mathbb{F}_q)$. Each point of $AG(4, \mathbb{F}_q)$ is denoted by (x_1, x_2, x_3, x_4) with $x_i \in \mathbb{F}_q$, $1 \le i \le 4$. Let V be a vector subspace of $AG(4, \mathbb{F}_q)$. A translation $\alpha + V$ of a k-dimensional vector space V, where α is any point of $AG(4, \mathbb{F}_q)$, is called a k-dimensional flat. The flats of dimension 1, 2, and 3 are called lines, planes, and superplanes, respectively. Two flats $\alpha + V_1$ and $\beta + V_2$ are called parallel if one of the subspaces V_1, V_2 is contained in the other one. Two parallel flats $\alpha + V$ and $\beta + V$ coincide if $\alpha - \beta \in V$; thus there are a total of $q^4/|V|$ parallel flats with the same dimension in the parallel group containing the flat $\alpha + V$.

Each line $\alpha + V$ with a 1-dimensional subspace V is the set of all points which satisfy a system of linear equations

$$a_1x_1 + b_1x_2 + c_1x_3 + d_1x_4 = g_1$$
$$a_2x_1 + b_2x_2 + c_2x_3 + d_2x_4 = g_2$$
$$a_3x_1 + b_3x_2 + c_3x_3 + d_3x_4 = g_3$$

where the rank of its coefficient matrix is 3. The subspace V is the set of all solutions of its corresponding system of linear homogeneous equations. Each line has q points on it. It is easy to see that any two different points α_1, α_2 determine a unique line

$$\alpha_1 + h(\alpha_1 - \alpha_2), \quad h \in AG(4, \mathbb{F}_q)$$

passing through them.

Each plane $\alpha + V$ with a 2-dimensional subspace V is the set of all points which satisfy a system of linear equations

$$a_1x_1 + b_1x_2 + c_1x_3 + d_1x_4 = g_1$$
$$a_2x_1 + b_2x_2 + c_2x_3 + d_2x_4 = g_2$$

where the rank of its coefficient matrix is 2. The subspace V is the set of all solutions of its corresponding system of linear homogeneous equations. Each plane has q^2 points on it. It is easy to see that any three different points α_1, α_2, α_3, located not on one line, determine a unique plane

$$\alpha_1 + h_1(\alpha_2 - \alpha_1) + h_2(\alpha_3 - \alpha_1), \quad h_1, h_2 \in AG(4, \mathbb{F}_q)$$

passing through them.

Each superplane $\alpha + V$ with a 3-dimensional subspace V is the set of all points which satisfy a linear equation

$$ax_1 + bx_2 + cx_3 + dx_4 = g$$

The subspace V is the set of all solutions of its corresponding linear homogeneous equation. Each superplane has q^3 points on it. It is easy to see that any four different points α_1, α_2, α_3, α_4, located not on one plane, determine a unique superplane

$$\alpha_1 + h_1(\alpha_2 - \alpha_1) + h_2(\alpha_3 - \alpha_1) + h_3(\alpha_4 - \alpha_1), \quad h_1, h_2, h_3 \in AG(4, \mathbb{F}_q)$$

passing through them.

LEMMA 4.1

(i) *Two flats* $\alpha + V_1$ *and* $\beta + V_2$ *intersect if and only if* $\alpha - \beta \in V_1 + V_2$, *and when they interest* $dim([\alpha + V_1] \cap [\beta + V_2]) = dim(V_1 \cap V_2)$.

(ii) *If one plane* $\alpha + V_1$ *and one superplane* $\beta + V_2$ *intersect, then* $V_1 \subset V_2$ *or* $(\alpha + V_1) \cap (\beta + V_2)$ *is a line.*

(iii) *If two superplanes* $\alpha + V_1$ *and* $\beta + V_2$ *intersect, then they coincide or* $(\alpha + V_1) \cap (\beta + V_2)$ *is a plane.*

PROOF   Two flats $\alpha + V_1$ and $\beta + V_2$ intersect if and only if there exist $v_1 \in V_1$ and $v_2 \in V_2$ such that $\alpha + v_1 = \beta + v_2$, the first conclusion of (i) is true. Put $\gamma = \alpha + v_1 = \beta + v_2$, then $\alpha + V_1 = \gamma + V_1, \beta + V_2 = \gamma + V_2$, hence $(\alpha + V_1) \cap (\beta + V_2) = \gamma + (V_1 \cap V_2)$, the second conclusion of (i) holds. If $dim V_1 = 2$ and $dim V_2 = 3$, then $V_1 \cap V_2$ or $dim(V_1 \cap V_2) = 1$, while if $dim V_1 = dim V_2 = 3$, then $V_1 = V_2$ or $dim(V_1 \cap V_2) = 2$, (ii) and (iii) follow from (i). $\qquad \Box$

Scheme B

Fix a line L_0 in $AG(4, \mathbb{F}_q)$. Take the points on L_0 as source states. Each superplane intersecting L_0 at a unique point is taken as a possible message. The decoding rules of the receiver are the lines neither intersecting L_0 nor parallel to L_0. The transmitter takes the planes which neither intersect L_0 nor parallel L_0 as encoding rules. An encoding rule e is valid under the decoding rule f if f is on e. Under an encoding rule e, a source state s is encoded into the superplane m passing through e and s. Under a decoding rule f, a message m is accepted by the receiver if f is on m, and is decoded into the intersecting point of m with L_0. Since any possible message has a unique intersecting point with the line L_0, this is a Cartesian A^2-code.

There are q points on each line; we have $|\mathscr{S}| = q$. Without loss of the generality we may assume that line L_0 is defined by $x_1 = x_2 = x_3 = 0$, i.e.,

$$L_0 = \{(0, 0, 0, x) \mid x \in \mathbb{F}_q\}.$$

A superplane defined by

$$ax_1 + bx_2 + cx_3 + dx_4 = g$$

has a unique intersecting point with L_0 if and only if $d \neq 0$, and the intersecting point is $(0, 0, 0, g/d)$. (When $d = g = 0$ the line L_0 is contained in the superplane while $d = 0, g \neq 0$ the line L_0 has no common point with the superplane.) We may assume that $d = 1$ when $d \neq 0$, hence, each possible message can be denoted by

$$ax_1 + bx_2 + cx_3 + x_4 = g. \tag{4.33}$$

This shows that $|\mathcal{M}| = q^4$. For any fixed source state $s = (0, 0, 0, g)$ there are q^3 messages which can represent it, i.e., $|\mathcal{M}(s)| = q^3$ for any $s \in \mathcal{S}$. Note that any line $\{(a_1, a_2, a_3, h), h \in \mathbb{F}_q\}$ parallel to L_0 is not contained in the superplane in Equation (4.33).

Given a decoding rule f and a source state s, let α_1, α_2 be two fixed points on f and $\alpha_3 = s$. Since L_0 neither intersects f nor parallels f, L_0 is not contained in the plane spanned by $\alpha_1, \alpha_2, \alpha_3$. There are $q^4 - q^2$ choices of the fourth point α_4 located outside the plane spanned by $\alpha_1, \alpha_2, \alpha_3$, and there are $q^3 - q^2$ choices of the fourth point α_4 in a superplane containing f and s, hence, the number of superplanes containing f and s is

$$\frac{q^4 - q^2}{q^3 - q^2} = q + 1.$$

When $\alpha_4 \in L_0$ we obtain a superlpane containing f, s, and L_0 which is unique. Therefore,

$$c = |\mathcal{M}(f, s)| = q$$

for any $f \in \mathcal{E}_R$ and $s \in \mathcal{S}$. Furthermore,

$$|\mathcal{M}(f)| = ck = q^2$$

for any $f \in \mathcal{E}_R$.

The total number of lines in $\mathbb{F}_q^4$ is

$$\frac{q^4(q^4 - 1)}{q(q - 1)} = q^3(q^2 + 1)(q + 1);$$

the total number of lines passing through one fixed point is

$$\frac{q^4 - 1}{q - 1} = (q^2 + 1)(q + 1);$$

hence, the total number of lines intersecting L_0 at one point is

$$q((q^2 + 1)(q + 1) - 1) = q^4 + q^3 + q^2.$$

The total number of lines parallel to L_0 is

$$q^4/q = q^3;$$

therefore, the total number of lines neither intersecting L_0 nor parallel to L_0 is

$$|\mathscr{E}_R| = q^3(q^2+1)(q+1) - (q^4+q^3+q^2) - q^3$$
$$= q^6 + q^5 - q^3 - q^2.$$

The total number of planes in $\mathbb{F}_q^4$ is

$$\frac{q^4(q^4-1)(q^4-q)}{q^2(q^2-1)(q^2-q)} = q^6 + q^5 + 2q^4 + q^3 + q^2;$$

the total number of planes passing through one fixed point is

$$\frac{(q^4-1)(q^4-q)}{(q^2-1)(q^2-q)} = q^4 + q^3 + 2q^2 + q + 1;$$

and the total number of planes passing through one fixed line is

$$\frac{q^4-q}{q^2-q} = q^2 + q + 1.$$

Hence, the total number of planes intersecting L_0 at one point is

$$q(q^4+q^3+2q^2+q+1 - (q^2+q+1)) = q^5 + q^4 + q^3.$$

Each plane parallel to L_0 is parallel to one plane passing through L_0 and the total number of planes parallel to one fixed plane is $q^4/q^2 = q^2$ as pointed out above,; hence, the total number of planes parallel to L_0 is

$$q^2(q^2+q+1) = q^4 + q^3 + q^2.$$

It follows that the total number of planes neither intersecting L_0 nor parallel to L_0 is

$$|\mathscr{E}_T| = q^6 + q^5 + 2q^4 + q^3 + q^2 - (q^5+q^4+q^3) - (q^4+q^3+q^2)$$
$$= q^6 - q^3.$$

Given an encoding rule $e = \alpha + V$, which is a plane neither intersecting L_0 nor parallel to L_0 (this implies that L_0 is not contained in V), any line in e neither intersects L_0 nor is parallel to L_0. Hence,

$$|\mathscr{E}_R(e)| = \frac{q^2(q^2-1)}{q(q-1)} = q^2 + q$$

for any $e \in \mathscr{E}_T$. Furthermore, we have

$$|\mathscr{E}_T(f)| = \frac{|\mathscr{E}_T||\mathscr{E}_R(e)|}{|\mathscr{E}_R|} = q^2$$

for any $f \in \mathscr{E}_R$.

Now we show that Scheme B satisfies the last three conditions of Theorem 4.1.

For a given encoding rule e and a message $m \in \mathscr{M}'(e)$, it implies that e is not contained in m. By (ii) of Lemma 4.1, $e \cap m$ is a line $f \in \mathscr{E}_R(e)$ and $m \in \mathscr{M}'_f(e)$. Such decoding rule f is unique. Hence, the pair $\{\mathscr{M}'(e), [\mathscr{M}'_f(e) \mid f \in \mathscr{E}_R(e)]\}$ is a $1 - (q^4 - q, q^2 + q, q^2 - q; 1, 0)$ design.

For a given decoding rule f and a message $m \in \mathscr{M}(f)$, it implies that $f \subset m$ and $m \cap L_0 = s$. The total number of planes e such that $f \subset e \subset m$ is

$$\frac{q^3 - q}{q^2 - q} = q + 1,$$

among them there is one plane passing through f and s. Hence, there are q encoding rules $e \in \mathscr{E}_T(f)$ such that $m \in \mathscr{M}(e)$. Let m_1, m_2 be two messages in $\mathscr{M}(f)$ with $m_1 \cap L_0 \neq m_2 \cap L_0$. If m_1 and m_2 intersect, then $m_1 \cap m_2$ is an encoding rule e (note there is no line in m_1 or m_2 which is parallel to L_0, so $m_1 \cap m_2$ neither intersects nor is parallel to L_0) such that $e \in \mathscr{E}_T(f)$ and $m_1, m_2 \in \mathscr{M}(e)$. Such encoding rule e is unique. This shows that the pair $\{\mathscr{M}(f), [\mathscr{M}(e) \mid e \in \mathscr{E}_T(f)]\}$ is a SPBD $2 - (q^2, q^2, q; q, 1, 0)$ design.

For a given message m, any line in m, which does not intersect $m \cap L_0$, is a decoding rule; hence, there are

$$\frac{q^3(q^3 - 1)}{q(q - 1)} - \frac{q^3 - 1}{q - 1} = q^4 + q^3 - q - 1$$

decoding rules f such that $m \in \mathscr{M}(f)$. For any two messages m_1, m_2 with $m_1 \cap L_0 \neq m_2 \cap L_0$, if $m_1 \cap m_2 \neq \emptyset$, then $e = m_1 \cap m_2$ is an encoding rule by (iii) of Lemma 4.1 (note that $m_1 \cap m_2$ neither intersects L_0 nor is parallel to L_0). All

$$\frac{q^2(q^2 - 1)}{q(q - 1)} = q^2 + q$$

lines in e are decoding rules. This proves that the pair $\{\mathscr{M}, [\mathscr{M}(f) \mid f \in \mathscr{E}_R]\}$ is a RSPBD $2 - (q^4, q^6 + q^5 - q^3 - q^2, q, q; q^4 + q^3 - q - 1, q^2 + q, 0)$ design.

We can find that

$$P_{O_0} = \frac{1}{q^2}, \quad P_{O_1} = \frac{q}{q^3 - 1}$$

$$P_{R_0} = P_{R_1} = \frac{1}{q}$$

$$P_T = \frac{1}{q^2 + q}$$

for Scheme B, which is a perfect Cartesian A^2-code of Type II.

Scheme B does not satisfy the assumption of Corollary 4.7. Let $s_1 = (0, 0, 0, g_1)$, $s_2 = (0, 0, 0, g_2)$, $(g_1 \neq g_2)$ be two source states and $m_1 \in \mathscr{M}(s_1)$,

$m_2 \in \mathcal{M}(s_2)$. Suppose that m_1 and m_2 are defined by

$$a_1 x_1 + b_1 x_2 + c_1 x_3 + x_4 = g_1$$

and

$$a_2 x_1 + b_2 x_2 + c_2 x_3 + x_4 = g_2$$

respectively. If $(a_1, b_1, c_1) = (a_2, b_2, c_2)$, then m_1 and m_2 do not intersect and $|\mathcal{E}_T(m_1, m_2)| = 0$, otherwise we have $|\mathcal{E}_T(m_1, m_2)| = 1$. This also implies that $|\mathcal{E}_T| = q^6 - q^3$.

4.4   Combinatorial Bounds of A^2-Codes

We consider the combinatorial bounds of P_{O_r}, P_{R_r}, and P_T in this section. Assume at first that for any $f \in \mathcal{E}_R$ and $s \in \mathcal{S}$ there exists a constant c such that

$$|\mathcal{M}(f, s)| = c.$$

It deduces that

$$|\mathcal{M}(f)| = kc$$

for any $f \in \mathcal{E}_R$.

PROPOSITION 4.6

$$P_{O_r} \geq \frac{c(k - r)}{v - r}$$

for $0 \leq r < k$. The equality holds if and only if

$$P(m|m^r) = \frac{c(k - r)}{v - r}$$

for any $m^r = (m_1, \cdots m_r) \in \overline{\mathcal{M}_R^r}$ and $m \in \mathcal{M}$ with $m \neq m_i (1 \leq i \leq r)$.

PROOF   Recall that $P(m|m^r)$ is the probability of the event that the message m is acceptable by the receiver given m^r has been accepted where $m \neq m_i (1 \leq i \leq r)$, and

$$P(m|m^r) = \sum_{f \in \mathcal{E}_R(m^r * m)} p(f|m^r).$$

We have

$$(v - r) \max_{m \in \mathcal{M}} P(m|m^r) \geq \sum_{m \neq m_i} P(m|m^r)$$

$$= \sum_{m \neq m_i} \sum_{f \in \mathcal{E}_{lt}(m'*m)} p(f|m^r)$$

$$= \sum_{f \in \mathcal{E}_{lt}(m^r)} p(f|m^r) \sum_{m:f(m) \neq f(m_i)} 1$$

$$= c(k - r),$$

hence,

$$(v - r)P_{O_r} = (v - r) \sum_{m' \in \mathcal{M}_{lt}^r} p(m^r) \max_{m \in \mathcal{M}} P(m|m^r)$$

$$\geq (kc - rc) \sum_{m' \in \mathcal{M}_{lt}^r} p(m^r)$$

$$= c(k - r).$$

Thus, the inequality

$$P_{O_r} \geq \frac{c(k - r)}{v - r}$$

is proved. The equality holds if and only if

$$P(m|m^r) = \frac{c(k - r)}{v - r}$$

for any $m^r \in \overline{\mathcal{M}_R^r}$ and any $m \in \mathcal{M}$ with $m \neq m_i (1 \leq i \leq r)$. □

PROPOSITION 4.7

$$P_{R_r} \geq \frac{1}{c}$$

for $0 \leq r < k$. The equality holds if and only if

$$P(m|f, m^r) = \frac{1}{c}$$

for any $f \in \mathcal{E}_R$, $m^r = (m_1, \cdots, m_r) \in \mathcal{M}^r(f)$ and $m \in \mathcal{M}(f)$ such that $f(m_i)(0 \leq i \leq r)$ and $f(m)$ are pairwise distinct.

PROOF Recall that $P(m|f, m^r)$ is the probability of the event that m is valid of the encoding rule e used by the transmitter given $f \in \mathcal{E}_R$ and $m^r = (m_1, \cdots, m_r) \in \mathcal{M}^r(e)$ such that $f(m_i)(0 \leq i \leq r)$ and $f(m)$ are pairwise distinct, and

$$P(m|f, m^r) = \sum_{e \in \mathcal{E}_T(f, m'*m)} p(e|f, m^r).$$

We have

$$(c(k-r)) \max_m P(m|f, m^r) \geq \sum_{m \in \mathcal{M}(f), f(m) \neq f(m_i)} P(m|f, m^r)$$

$$= \sum_{e \in \mathcal{E}_T(f, m^r)} p(e|f, m^r) \sum_{m \in \mathcal{M}(e), m \neq m_i} 1$$

$$= k - r,$$

hence,

$$(c(k-r)) P_{R_r}$$
$$= c(k-r) \sum_{f \in \mathcal{E}_R} p(f) \sum_{m^r \in \mathcal{M}^r(f)} p(m^r|f) \max_{m \in \mathcal{M}(f), f(m) \neq f(m_i)} P(m|f, m^r))$$
$$\geq (k-r) \sum_{f \in \mathcal{E}_R} p(f) \sum_{m^r \in \mathcal{M}^r(f)} p(m^r|f)$$
$$= k - r.$$

Thus, the inequality

$$P_{R_r} \geq \frac{k-r}{c(k-r)} = \frac{1}{c}$$

is proved. The equality holds if and only if

$$P(m|f, m^r) = \frac{1}{c}$$

for any $f \in \mathcal{E}_R$, $m^r = (m_1, \cdots, m_r) \in \mathcal{M}^r(f)$ and $m \in \mathcal{M}(f)$ such that $f(m_i) (0 \leq i \leq r)$ and $f(m)$ are pairwise distinct. □

PROPOSITION 4.8

$$P_T \geq \frac{k(c-1)}{v-k}.$$

The equality holds if and only if

$$P(m'|e) = \frac{k(c-1)}{v-k}$$

for any $e \in \mathcal{E}_T$ and $m' \in \mathcal{M}'(e)$.

PROOF   Recall that $P(m'|e)$ is the probability of the event that the message m' is acceptable by the receiver given $e \in \mathcal{E}_T$, and

$$P(m'|e) = \sum_{f \in \mathcal{E}_R(e, m')} p(f|e).$$

We have

$$(v-k)\max_{m'} P(m'|e) \geq \sum_{m'} P(m'|e)$$

$$= \sum_{f \in \mathscr{E}_R(e)} p(f|e) \sum_{m' \in \mathscr{M}'_f(e)} 1$$

$$= (kc - k) \sum_{f \in \mathscr{E}_R(e)} p(f|e)$$

$$= k(c-1),$$

hence,

$$(v-k)P_T = (v-k) \sum_{e \in \mathscr{E}_T} p(e) \max_{m'} P(m'|e)$$

$$\geq k(c-1).$$

Thus, the inequality

$$P_T \geq \frac{k(c-1)}{v-k}$$

is proved, and the equality holds if and only if

$$P(m'|e) = \frac{k(c-1)}{v-k}$$

for any $e \in \mathscr{E}_T$ and $m' \in \mathscr{M}'(e)$. ⬜

In order to consider the combinatorial structure of the perfect authentication schemes with arbitration whose P_{O_r}, P_{R_R}, and P_T achieve combinatorial bounds, we should introduce the definition of restricted t-designs.

DEFINITION 4.3

Let $v, b, k, c, \mu, t, (t < k)$ be positive integers. A t-design $(\mathscr{M}, E)$ is called a restricted $t - (v, b, k, c, \mu)$ design if $|\mathscr{M}| = v$, $\mathscr{E} = (\mathscr{E}_1, \cdots, \mathscr{E}_b)$, each block $\mathscr{E}_i$ is divided into k groups, and each group has c points of $\mathscr{M}$. Any t-subset of $\mathscr{M}$ occurs in μ blocks in such a way that each point of the t-subset occupies one group.

REMARK 4.2 A restricted $t - (v, b, k, c, \mu)$ design is also a restricted $r - (v, b, k, c, \mu_r)$ design for any $1 \leq r \leq t-1$ where

$$\mu_r = \frac{(v-r) \cdots (v-t+1)}{c^{t-r}(k-r) \cdots (k-t+1)} \cdot \mu. \tag{4.34}$$

(Note that $bkc = v\mu_1$, which will be used later on.) ⬜

THEOREM 4.2

Let $\mathscr{A} = (\mathscr{S}, \mathscr{M}, \mathscr{E}_R, \mathscr{E}_T, p_S, p_{E_R}, p_{E_T})$ be an authentication scheme with arbitration, and t be an integer with $t < k$. The following conditions

(a) $P_{O_r} = \frac{c(k-r)}{v-r}, \quad 0 \le r \le t - 1$;

(b) $P_{R_r} = \frac{1}{c}, \quad 0 \le r \le t - 1$;

(c) $P_T = \frac{k(c-1)}{v-k}$;

(d) $|\mathscr{E}_T| = \frac{v(v-1)\cdots(v-t+1)}{k(k-1)\cdots(k-t+1)}$;

(e) $|\mathscr{E}_R| = \frac{v(v-1)\cdots(v-t+1)(v-k)}{c^t(c-1)k^2(k-1)\cdots(k-t+1)}$

are satisfied if and only if

(i) $E_T, E_R, E_T(f)(\forall f \in \mathscr{E}_R)$ and $E_R(e)(\forall e \in \mathscr{E}_T)$, have uniform probability distribution.

(ii) For any given $m^r \in \overline{\mathscr{M}_R^r}(1 \le r \le t)$, the probability $p(S^r = f[m^r])$ is constant for all $f \in \mathscr{E}_R(m^r)$.

(iii) For any given $e \in \mathscr{E}_T$ the pair

$$(\mathscr{M}'(e), \{\mathscr{M}_f'(e) | f \in \mathscr{E}_R(e)\})$$

is a $1 - (v - k, b_1, k[c - 1], 1)$ design where $v = |\mathscr{M}|, \ k = |\mathscr{S}|, \ b_1 = |\mathscr{E}_R(e)| = (v-k)/k(c-1)$, c is a positive integer (in fact, $c = |\mathscr{M}(f,s)|$ for all $f \in \mathscr{E}_R, s \in \mathscr{S}$).

(iv) For any given $f \in \mathscr{E}_R$, the pair

$$(\mathscr{M}(f), \{\mathscr{M}(e) | e \in \mathscr{E}_T(f)\})$$

is a SPBD $t - (kc, b_2, k; \lambda_1, \cdots, \lambda_t, 0)$ where $\lambda_t = 1, \ \lambda_r = c^{t-r}, 1 \le r \le t - 1, b_2 = |\mathscr{E}_T(f)| = c^t$.

(v) The pair

$$(\mathscr{M}, \{\mathscr{M}(f) | f \in \mathscr{E}_R\})$$

is a restricted $t - (v, b_3, k, c, \mu)$ design, where $\mu = P_T^{-1}, b_3 = |\mathscr{E}_R|$.

PROOF   Assume that the conditions (a) through (e) in the theorem hold. Then it implies that

$$|\mathscr{E}_T| = (P_{O_0} P_{O_1} \cdots P_{O_{t-1}} P_{R_0} P_{R_1} \cdots P_{R_{t-1}})^{-1},$$

and

$$|\mathscr{E}_R| = (P_{O_0} P_{O_1} \cdots P_{O_{t-1}} P_T)^{-1}.$$

The scheme $\mathscr{A}$ is t-fold perfect, and the conditions (i) through (v) of Theorem 4.1 hold. Now we only need to show that the SPBD in items (iii) and (v) are (restricted) t-designs.

It deduces from condition (c) and Proposition 4.8 that $|\mathscr{E}_R(f, m')| > 0$ for any $e \in \mathscr{E}_T$ and $m' \in \mathscr{M}'(e)$, which means that

$$(\mathscr{M}'(e), \{\mathscr{M}'_f(e)|f \in \mathscr{E}_R(e)\})$$

is a $1 - (v - k, b_1, k[c - 1], 1)$ design where $v = |\mathscr{M}|$, $k = |\mathscr{S}|$, $b_1 = |\mathscr{E}_R(e)|$, $c = |\mathscr{M}(f, s)|$ for all $f \in \mathscr{E}_R, s \in \mathscr{S}$.

It can be shown that $|\mathscr{E}_R(m^r * m)| > 0$, $0 \le r \le t - 1$ for any $m^r = (m_1, \cdots, m_r) \in \overline{\mathscr{M}_R^r}$ and $m \in \mathscr{M}$ with $m \ne m_i (1 \le i \le r)$ from the condition (a) and Proposition 4.6. Since $\mathscr{M} = \overline{\mathscr{M}_R^1}$, it deduces by induction on $2 \le r \le t - 1$ that $|\mathscr{E}_R(m^t)| > 0$ for any $m^t \in \mathscr{M}^t$. Therefore, the pair

$$(\mathscr{M}, \{\mathscr{M}(f)|f \in \mathscr{E}_R\})$$

is a restricted $t - (v, b_3, k, c, \mu)$ design, where $\mu = P_T^{-1}, b_3 = |\mathscr{E}_R|$.

Conversely, suppose that the conditions (i) through (v) in the theorem hold. Let $m^r = (m_1, \cdots, m_r) \in \overline{\mathscr{M}_R^r}$, then

$$p(f|m^r) = \frac{p(f(m^r))p(f)}{p(m^r)}$$

does not depend on $f \in \mathscr{E}_T(m^r)$ by (i) and (ii). Hence,

$$\begin{aligned} p(f|m^r) &= \frac{1}{|\mathscr{E}_R(m^r)|} \\ &= \frac{c^{t-r}(k - r) \cdots (k - t + 1)}{(v - r) \cdots (v - t + 1)\mu} \end{aligned}$$

by Equation (4.34), and

$$P(m|m^r) = \frac{|\mathscr{E}_R(m^r * m)|}{|\mathscr{E}_R(m^r)|} = \frac{c(k - r)}{v - r}$$

for any $m \in \mathscr{M}$ with $m \ne m_i, 1 \le i \le r$. The condition (a) is satisfied from Proposition 4.6.

For any $f \in \mathscr{E}_R$ and $m^r = (m_1, \cdots, m_r) \in \mathscr{M}^r(f)$ with $\mathscr{E}_T(f, m^r) \ne \emptyset$,

$$p(e|f, m^r) = \frac{p(e)p(f|e)p(f(m^r))}{p(f, m^r)}$$

does not depend on $e \in \mathscr{E}_T(f, m^r)$ by (i), hence,

$$p(e|f, m^r) = \frac{1}{|\mathscr{E}_T(f, m^r)|} = \frac{1}{c^{t-r}}$$

by (iv) and

$$P(m|f, m^r) = \frac{|\mathscr{E}_T(f, m^r * m)|}{|\mathscr{E}_T(f, m_r)|} = \frac{1}{c}$$

for any $m \in \mathscr{M}(f)$ with $f(m) \neq f(m_i), 1 \leq i \leq r$. The condition (b) is satisfied from Proposition 4.7.

Given any $e \in \mathscr{E}_T$ and $m' \in \mathscr{M}'(e)$, we have $|\mathscr{E}_R(e, m')| = 1$ and

$$P(m'|e) = \frac{1}{|\mathscr{E}_R(e)|} = \frac{1}{b_1} = \frac{k(c-1)}{v - k}$$

by (iii). Hence, the condition (c) is satisfied from Proposition 4.8. It deduces also that

$$\mu = \frac{v - k}{k(c - 1)}.$$

It follows from (v) and Equation (4.34) that

$$|\mathscr{E}_R| = b_3 = \frac{v}{kc} \cdot \mu_1$$
$$= \frac{v(v-1)\cdots(v-t+1)(v-k)}{c^t(c-1)k^2(k-1)\cdots(k-t+1)}.$$

This is the condition (e). Notice that $b_2 = |\mathscr{E}_T(f)| = c^t$, hence,

$$|\mathscr{E}_T| = \frac{|\mathscr{E}_R||\mathscr{E}_T(f)|}{\mathscr{E}_R(e)|}$$
$$= \frac{v(v-1)\cdots(v-t+1)}{k(k-1)\cdots(k-t+1)}.$$

This completes the proof of Theorem 4.2.   ⬚

We call the schemes which satisfy the conditions given in Theorem 4.2 as perfect A^2-schemes with combinatorial bounds.

4.5   Comments

All the results of Sections 4.1 and 4.2 are due to Pei and Li [33], and Pei et al. [31]. The proofs of Propositions 4.1, 4.2, and 4.3 are similar to the proof of Proposition 3.1. Johansson [16] also obtained these bounds.

Corollary 4.7 is due to the author. Scheme A in Section 4.3 was constructed by Johansson [16], and Scheme B was constructed by Hu [15]. Two constructions in Exercises 4.2 and 4.3 were also presented by Hu in the same paper.

Theorem 4.2 is due to the author.

4.6 Exercises

4.1 Prove that in the space $PG(3, \mathbb{F})$, any two different lines may intersect at one point or not intersect. A line and a plane may intersect at one point or the line is contained in the plane. Any two different planes always intersect at one line.

4.2 Fix a line L_0 in $\mathbb{F}_q^3$. Take the points on L_0 as source states. Each plane intersecting L_0 at a unique point is regarded as a possible message. The receiver takes points f not on L_0 as decoding rules. The encoding rules of the transmitter are the lines e which neither intersect L_0 nor parallel L_0. An encoding rule e is valid under a decoding rule f if f is on e. Under an encoding rule e, a source state is encoded into the plane passing through e and s. Under a decoding rule f, a message m is accepted by the receiver if f is on m, and decoded into the intersecting point of L_0 and m. Prove that this code is a perfect Cartesian A^2-code of Type II.

4.3 Fix a line L_0 in $PG(4, \mathbb{F}_q)$. Take the points on L_0 as source states. A 3-dimensional projective subspace in $PG(4, \mathbb{F}_q)$ is called a superplane. Each superplane intersecting L_0 at a unique point is taken as a possible message. The decoding rules f of the receiver are the lines not intersecting L_0. The transmitter takes the planes which do not intersect L_0 as encoding rules e. An encoding rule e is valid under a decoding rule f if f is on e. Under an encoding rule e, a source state s is encoded into the unique superplane m passing through e and s. Under a decoding rule f, a message m is accepted by the receiver if f is on m, and is decoded into the intersecting point of L_0 and m. Prove that this code is a perfect Cartesian A^2-code of Type I.

4.4 Prove the equality in Equation (4.34).

Chapter 5

A-Codes Based on Rational Normal Curves

We have shown in the previous chapters that construction of perfect authentication schemes is reduced to construction of strong partially balanced design (SPBD). As we know from literature that there are mainly two kinds of known combinatorial designs, which are SPBD too: t-designs and orthogonal arrays [1]. Some special family of partially balanced incomplete block (PBIB [34]) design can provide SPBD with $t = 2$. In this chapter we construct a new family of SPBD by means of rational normal curves (RNC) over finite fields. Then we discuss the authentication schemes constructed based on this new SPBD.

5.1 SPBD Based on RNC

Let $\mathbb{F}_q$ be the finite field with q elements and $n \geq 2$ be a positive integer. Let $PG(n, \mathbb{F}_q)$ be the projective space of dimension n over $\mathbb{F}_q$. A point of $PG(n, \mathbb{F}_q)$ is denoted by $(x_0, x_1, \cdots, x_n)$ where $x_i \in \mathbb{F}_q$, $0 \leq i \leq n$ are not all zero. If λ is a nonzero element of $\mathbb{F}_q$, then $(\lambda x_0, \lambda x_1, \cdots, \lambda x_n)$ and $(x_0, x_1, \cdots, x_n)$ denote the same point of $PG(n, \mathbb{F}_q)$. Suppose that T is a nonsingular matrix over $\mathbb{F}_q$ of order $n + 1$, then T generates a one-to-one transformation of points in $PG(n, \mathbb{F}_q)$ defined by

$$PG(n, \mathbb{F}_q) \longrightarrow PG(n, \mathbb{F}_q)$$
$$(x_0, x_1, \cdots, x_n) \longmapsto (x_0, x_1, \cdots, x_n)T$$

It is called a *projective transformation* of $PG(n, \mathbb{F}_q)$. The group of all projective transformations of $PG(n, \mathbb{F}_q)$ is denoted by $PGL_{n+1}(\mathbb{F}_q)$. It is nothing but the factor group of the linear group $GL_{n+1}(\mathbb{F}_q)$ of order $n + 1$ over its subgroup $\{\lambda I_{n+1} \mid \lambda \neq 0\}$ where I_{n+1} is the identity matrix.

We define a curve $\mathcal{C}$ in $PG(n, \mathbb{F}_q)$ to be the image of the map

$$PG(1, \mathbb{F}_q) \longrightarrow PG(n, \mathbb{F}_q)$$
$$(x_0, x_1) \longmapsto (x_0^n, x_0^{n-1}x_1, \cdots, x_1^n). \tag{5.1}$$

$PG(1, \mathbb{F}_q)$ consists of the following $q + 1$ points:

$$\{(1, \alpha) : \alpha \in \mathbb{F}_q\} \cup \{(0, 1)\}.$$

Therefore, the curve $\mathcal{C}$ consists of the following $q + 1$ points:

$$\{(1, \alpha, \alpha^2, \cdots, \alpha^n) : \alpha \in \mathbb{F}_q\} \cup \{(0, 0, 0, \cdots, 0, 1)\} \tag{5.2}$$

It is easy to see that the $q + 1$ points in Equation (5.2) are all the solutions of the following system of homogeneous equations:

$$\begin{cases} X_i^2 - X_{i-1}X_{i+1} = 0, & 1 \le i \le n - 1 \\ X_1 X_{n-1} - X_0 X_n = 0. \end{cases}$$

We call the image of the curve $\mathcal{C}$ under any projective transformation a *rational normal curve* (or **RNC**, in short). In other words, a RNC is the image of a map

$$\begin{aligned} PG(1, \mathbb{F}_q) &\longrightarrow PG(n, \mathbb{F}_q) \\ (x_0, x_1) &\longmapsto (A_0(x_0, x_1), \cdots, A_n(x_0, x_1)) \end{aligned}$$

where $A_0(X_0, X_1), \cdots, A_n(X_0, X_1)$ is an arbitrary basis of the space of homogeneous polynomials of degree n over X_0, X_1. Obviously, there are $q+1$ points on each RNC of $PG(n, \mathbb{F}_q)$. We call m points $p_1, p_2, \cdots, p_m \in PG(n, \mathbb{F}_q)$, $p_i = (x_{i0}, x_{i1}, \cdots, x_{in})$, $1 \le i \le m$ linearly independent (or in general position) if the rank of the matrix $(x_{ij})_{1 \le i \le m, 0 \le j \le n}$ is m.

LEMMA 5.1
Any $n + 1$ ($n \le q$) points on a RNC in $PG(n, \mathbb{F}_q)$ are linearly independent.

PROOF   It is only necessary to prove the lemma for the curve $\mathcal{C}$ defined in Equation (5.1). By using Vandermonde determinant

$$\begin{vmatrix} 1 & \alpha_1 & \alpha_1^2 & \cdots & \alpha_1^n \\ 1 & \alpha_2 & \alpha_2^2 & \cdots & \alpha_2^n \\ \vdots & & & \ddots & \\ 1 & \alpha_{n+1} & \alpha_{n+1}^2 & \cdots & \alpha_{n+1}^n \end{vmatrix} = (-1)^{n(n+1)/2} \prod_{1 \le i < j \le n+1} (\alpha_i - \alpha_j)$$

and

$$\begin{vmatrix} 1 & \alpha_1 & \alpha_1^2 & \cdots & \alpha_1^n \\ 1 & \alpha_2 & \alpha_2^2 & \cdots & \alpha_2^n \\ & & & \ddots & \\ 0 & 0 & 0 & \cdots & 1 \end{vmatrix} = (-1)^{n(n+1)/2} \prod_{1 \le i < j \le n} (\alpha_i - \alpha_j),$$

the lemma can be proved easily.   ☐

LEMMA 5.2

Suppose that $q \geq n + 2$. For any $n + 3$ points in $PG(n, \mathbb{F}_q)$, if any $n + 1$ points among which are linearly independent, then there exists a unique RNC passing through these $n + 3$ points..

PROOF   After taking a projective transformation if necessary, we may assume that these $n + 3$ given points have the following form:

$$
\left\{
\begin{aligned}
p_1 &= (1, 0, \cdots, 0) \\
p_2 &= (0, 1, \cdots, 0) \\
&\ \ \vdots \\
p_{n+1} &= (0, 0, \cdots, 1) \\
p_{n+2} &= (1, 1, \cdots, 1) \\
p_{n+3} &= (v_0, v_1, \cdots, v_n)
\end{aligned}
\right.
\tag{5.3}
$$

Since any $n + 1$ points among them are linearly independent, we know that $v_0, v_1, \cdots, v_n$ are pairwise distinct nonzero elements in $\mathbb{F}_q$. Define a polynomial

$$
G(X_0, X_1) = \prod_{i=0}^{n} (X_0 - v_i^{-1} X_1)
$$

in X_0, X_1. Then the polynomials

$$
H_i(X_0, X_1) = \frac{G(X_0, X_1)}{X_0 - v_i^{-1} X_1}, \quad 0 \leq i \leq n
$$

form a basis for the space of homogeneous polynomials of degree n. Otherwise, if there were a linear relation

$$
\sum_{i=0}^{n} a_i H_i(X_0, X_1) = 0, \quad a_i \in \mathbb{F}_q,
$$

then substituting $(X_0, X_1) = (1, v_j)$ for $0 \leq j \leq n$ successively we could deduce that $a_j = 0$, $0 \leq j \leq n$. Thus, the RNC defined by the map

$$
\sigma : (x_0, x_1) \longmapsto (H_0(x_0, x_1), H_1(x_0, x_1), \ldots, H_n(x_0, x_1))
\tag{5.4}
$$

passes through the $n + 3$ points p_i, $1 \leq i \leq n + 3$, since $\sigma(1, v_i) = p_i$, $1 \leq i \leq n + 1$, $\sigma(1, 0) = p_{n+2}$ and $\sigma(0, 1) = p_{n+3}$. Conversely any RNC passing through the points p_i, $1 \leq i \leq n + 3$ can be written in the form of Equation (5.4). This completes the proof of the Lemma 5.2.   □

LEMMA 5.3

Let $n \geq 2$ be an integer and let $q \geq n+2$ be a prime power. Then the number of RNC in $PG(n, \mathbb{F}_q)$ is

$$q^{\frac{n(n+1)}{2}-1} \prod_{i=3}^{n+1} (q^i - 1).$$

PROOF   We consider such sets of points with the following property: each set consists of $n+3$ points that any $n+1$ points among them are linearly independent. Let $\mathcal{N}$ denote the number of such sets. Suppose $\{p_1, p_2, \cdots, p_{n+3}\}$ is such a set. There are totally $(q^{n+1} - 1)/(q-1)$ points in $PG(n, \mathbb{F}_q)$. The point p_1 can be any point of $PG(n, \mathbb{F}_q)$. For a fixed p_1, there are

$$\frac{q^{n+1} - 1}{q-1} - 1 = \frac{q(q^n - 1)}{q-1}$$

possible choices for p_2, as p_1 and p_2 are linearly independent. Generally, assume that $p_1, p_2, \cdots, p_i$, $1 \leq i \leq n$ have been chosen, the point p_{i+1} may have

$$\frac{q^{n+1} - 1}{q-1} - \frac{q^i - 1}{q-1} = \frac{q^i(q^{n+1-i} - 1)}{q-1}$$

choices, since there are $(q^i - 1)/(q-1)$ points which are linearly dependent on $p_1, p_2, \cdots, p_i$. Now we suppose that the points $p_1, p_2, \cdots, p_{n+1}$ have been chosen. By taking a projective transformation, we can assume that the points $p_1, p_2, \cdots, p_{n+1}$ are of the form in Equation (5.3). Hence, the point p_{n+2} is of the form $(a_0, \cdots, a_n)$ with $a_i \neq 0$, $0 \leq i \leq n$, so it has

$$\frac{(q-1)^{n+1}}{q-1} = (q-1)^n$$

choices. Finally, if the points $p_1, p_2, \cdots, p_{n+2}$ have been chosen, we can also assume that $p_1, \cdots, p_{n+1}$ are of the form in Equation (5.3). Then the point p_{n+3} is of the form $(b_0, \cdots, b_n)$ with $b_i \neq 0$, $b_i \neq b_j$, $0 \leq i < j \leq n$; it has

$$(q-2)(q-3) \cdots (q-n-1)$$

choices. Therefore, we have proved that

$$|\mathcal{N}| = \prod_{i=0}^{n} \frac{q^i(q^{n+1-i} - 1)}{q-1} \cdot (q-1)^n \prod_{i=2}^{n+1} (q-i) \cdot ((n+3)!)^{-1}$$

$$= q^{n(n+1)/2}((n+3)!)^{-1} \prod_{i=2}^{n+1} (q^i - 1)(q-i).$$

There are $q+1$ points on each RNC, hence, there are C_{q+1}^{n+3} sets from $\mathcal{N}$ on each RNC; here C_{q+1}^{n+3} is the binomial coefficient. The number of RNC in

$PG(n, \mathbb{F}_q)$ must be

$$\frac{\mathcal{N}}{C_{q+1}^{n+3}} = q^{n(n+1)/2-1} \prod_{i=3}^{n+1} (q^i - 1).$$

This completes the proof of Lemma 5.3. ⬚

Let $\mathcal{M}$ denote the set of all points in $PG(n, \mathbb{F}_q)$ and $\mathcal{E}$ the set of all RNC in $PG(n, \mathbb{F}_q)$. The following theorem will show that the pair $(\mathcal{M}, \mathcal{E})$ is a SPBD.

THEOREM 5.1
Let $t \geq 5$ be an integer, and let $q \geq t - 1$ be a prime power. There exists a SPBD $t - (v, b, k; \lambda_1, \cdots, \lambda_t, 0)$ where

$$v = (q^{t-2} - 1)/(q - 1),$$
$$k = q + 1,$$
$$b = q^{\frac{(t-2)(t-3)}{2} - 1} \prod_{i=3}^{t-2} (q^i - 1),$$

and

$$\lambda_1 = q^{\frac{(t-2)(t-3)}{2} - 1} \prod_{i=2}^{t-3} (q^i - 1),$$

$$\lambda_r = q^{\frac{(r+t-3)(t-2-r)}{2}} (q-1)^{r-2} \prod_{i=1}^{t-2-r} (q^i - 1) \prod_{i=1}^{r-2} (q - i), \quad 2 \leq r \leq t - 3,$$

$$\lambda_{t-2} = (q - 1)^{t-4} \prod_{i=1}^{t-4} (q - i),$$

$$\lambda_{t-1} = \prod_{i=2}^{t-3} (q - i),$$

$$\lambda_t = 1.$$

{Here we use the convention that $\prod_{i=1}^{r-2}(q - i) = 1$ if $r = 2$.}

PROOF Take $\mathcal{M} = PG(t - 3, \mathbb{F}_q)$ as the set of points, and all RNC in $PG(t - 3, \mathbb{F}_q)$ as the family of blocks $\mathcal{E}$. Consider any given t points of $PG(t - 3, \mathbb{F}_q)$; if any $t - 2$ points among them are linearly independent, then there exists a unique RNC passing through them. Otherwise there is no RNC passing through them. Therefore, $(\mathcal{M}, \mathcal{E})$ is a $t - (v, b, k; 1, 0)$.

Now we prove that $(\mathcal{M}, \mathcal{E})$ is a SPBD $t - (v, b, k; \lambda_1, \cdots, \lambda_t, 0)$. Given any $r (1 \leq r \leq t - 2)$ points of $PG(t - 3, \mathbb{F}_q)$, if they are linearly dependent then there is no RNC passing through them; if they are linearly independent then by the proof of Lemma 5.3 there are λ_r RNCs passing through them where

$$\lambda_1 = \prod_{i=1}^{t-3} \frac{q^{t-2}-q^i}{q-1} \cdot (q-1)^{t-3} \prod_{i=2}^{t-2}(q-i) \prod_{i=0}^{t-2}(q-i)^{-1}$$

$$= q^{\frac{(t-2)(t-3)}{2}-1} \prod_{i=2}^{t-3}(q^i-1),$$

$$\lambda_r = \prod_{i=r}^{t-3} \frac{q^{t-2}-q^i}{q-1} \cdot (q-1)^{t-3} \prod_{i=2}^{t-2}(q-i) \prod_{i=r-1}^{t-2}(q-i)^{-1}$$

$$= q^{\frac{(r+t-3)(t-2-r)}{2}}(q-1)^{r-2} \prod_{i=1}^{t-2-r}(q^i-1) \prod_{i=1}^{r-2}(q-i), \quad 2 \le r \le t-3,$$

$$\lambda_{t-2} = (q-1)^{t-3} \prod_{i=2}^{t-2}(q-i) \prod_{i=t-3}^{t-2}(q-i)^{-1}$$

$$= (q-1)^{t-4} \prod_{i=1}^{t-4}(q-i).$$

Finally, for any given $t-1$ points of $PG(t-3, \mathbb{F}_q)$, if there exist $t-2$ points that are linearly dependent among them, then there is no RNC passing through the given $t-1$ points; otherwise there are

$$\lambda_{t-1} = \prod_{i=2}^{t-2}(q-i)(q-t+2)^{-1} = \prod_{i=2}^{t-3}(q-i)$$

RNCs passing through the given $t-1$ points. This completes the proof of the theorem. □

5.2   A Family of Non-Cartesian Perfect A-Codes

Based on the SPBD constructed in the previous section, one new type of *non-Cartesian* perfect authentication code can be constructed.

Let the set of source states $\mathcal{S}$ be the set of all points on the curve $\mathcal{C}$, then $k = q+1$. It can also be represented by $\mathcal{S} = \mathbb{F}_q \cup \{\infty\}$. Let the set of messages $\mathcal{M}$ be the set of all points in $PG(t-3, \mathbb{F}_q)$, then $|\mathcal{M}| = (q^{t-2}-1)/(q-1)$. For a RNC C^*, let T^* be a projective transformation which carries C to C^* and define an encoding rule σ_{T^*} as follows:

$$\sigma_{T^*}(\alpha) = (1, \alpha, \cdots, \alpha^{t-2}, \alpha^{t-3})T^*, \quad \alpha \in \mathbb{F}_q$$
$$\sigma_{T^*}(\infty) = (0, 0, \cdots, 0, 1)T^*.$$

Receiving an $m \in \mathcal{M}$, the receiver decides to accept m if $m(T^*)^{-1}$ is on $\mathcal{C}$ (it is just the source state transmitted) or to reject it otherwise. The number of encoding rules is the number of RNCs, and, hence, from Lemma 5.3 it is equal to

$$|\mathcal{E}| = q^{\frac{(t-2)(t-3)}{2}-1} \prod_{i=3}^{t-2}(q^i - 1)$$

Assume that the random variables E and S^r, $1 \le r \le t$, satisfy the conditions given in Theorem 3.1. Then we have

$$P_0 = \frac{\lambda_1}{|\mathcal{E}|} = \frac{q^2 - 1}{q^{t-2} - 1}$$

$$P_1 = \frac{\lambda_2}{\lambda_1} = \frac{q - 1}{q^{t-3} - 1}$$

$$P_r = \frac{\lambda_{r+1}}{\lambda_r} = \frac{(q-1)(q-r+1)}{q^{t-2} - q^r}, \quad 2 \le r \le t - 4$$

$$P_{t-3} = \frac{\lambda_{t-2}}{\lambda_{t-3}} = \frac{q - t + 4}{q^{t-3}}$$

$$P_{t-2} = \frac{\lambda_{t-1}}{\lambda_{t-2}} = \frac{q - t + 3}{(q-1)^{t-3}}$$

$$P_{t-1} = \frac{\lambda_t}{\lambda_{t-1}} = \prod_{i=2}^{t-3}(q-i)^{-1}.$$

We consider the encoding rules of the authentication schemes constructed above. Let $PO_{n+1}(\mathbb{F}_q)$ denote the subgroup of $PGL_{n+1}(\mathbb{F}_q)$ consisting of the projective transformations which carry the curve C into itself. Two projective transformations T_1 and T_2 carry the curve C into the same RNC if and only if they belong to the same coset of $PO_{n+1}(\mathbb{F}_q)$ in $PGL_{n+1}(\mathbb{F}_q)$. Thus, we can take any representation system for cosets of $PO_{n+1}(\mathbb{F}_q)$ in $PGL_{n+1}(\mathbb{F}_q)$ as the set of encoding rules. To find such a representation system is not easy. We will give such a representation system for the case when $t = 5$ in Sections 5.3 and 5.4 for odd q and even q, respectively. Instead of choosing such a representation system we can also use the following way, which needs a certain amount of calculation, to choose encoding rules. As mentioned above, the transmitter and receiver can determine an encoding rule by randomly choosing a nonsingular matrix T of order $n + 1$ over $\mathbb{F}_q$. Supposing that the matrices T_1 and T_2 have been randomly chosen successively, it is necessary to check whether T_1 and T_2 belong to the same coset of $PO_{n+1}(\mathbb{F}_q)$, i.e., to check whether $T_1 T_2^{-1}$ is an element of $PO_{n+1}(\mathbb{F}_q)$. This can be done by checking whether the point set in Equation (5.2) is carried into itself by the projective transformation $T_1 T_2^{-1}$. Particularly, it is easy to see whether the point $(0, \cdots, 0, 1)T_1 T_2^{-1}$ belongs to the set in Equation (5.2). If it is the case, we can try another matrix T_2 immediately. The probability of the event that $(0, \cdots, 0, 1)T_1 T_2^{-1}$ belongs to the set in Equation (5.2) is small (see below).

In this way we can successively choose several matrices T such that any two of them belong to the different cosets of $PO_{n+1}(\mathbb{F}_q)$. This work can be done in advance of the transmission. Since

$$|PGL_{n+1}(\mathbb{F}_q)| = q^{n(n+1)/2} \prod_{i=2}^{n+1}(q^i - 1)$$

we have, by Lemma 5.3, that

$$|PO_{n+1}(\mathbb{F}_q)| = \frac{|PGL_{n+1}(\mathbb{F}_q)|}{|\mathscr{E}|} = q(q^2 - 1).$$

Therefore, the probability of the event that two successively chosen matrices belong to the same coset of $PO_{n+1}(\mathbb{F}_q)$ is

$$\frac{|PO_{n+1}(\mathbb{F}_q)| - 1}{|PGL_{n+1}(\mathbb{F}_q)| - 1} \sim q^{-(n-1)(n+3)}.$$

The probability of the event that the $(n+1)^{st}$ row of $T_1 T_2^{-1}$ is a point of the curve C is

$$\frac{(q+1)(q-1)}{q^{n+1} - 1} \sim q^{-(n-1)}.$$

Let us come back to the problem: to find a representation system for cosets of $PO_{n+1}(\mathbb{F}_q)$ in $PGL_{n+1}(\mathbb{F}_q)$. At first we give the composition of the group $PO_{n+1}(\mathbb{F}_q)$. For $0 \le i, k \le n$, and $t_1, t_2 \in \mathbb{F}_q$, define

$$b_{i,k} = \begin{cases} C_k^i, & i \le k \\ 0, & i > k \end{cases}$$

and

$$c_{i,k} = \sum_{j=\max(0,i-k)}^{\min(i,n-k)} C_{n-k}^j C_k^{i-j} t_1^{k-(i-j)} t_2^{i-j}$$

where C_k^i is the binomial coefficient. For any elements $a \in \mathbb{F}_q^*$ (the set of nonzero element of $\mathbb{F}_q$) and $t \in \mathbb{F}_q$, define two matrices:

$$T_{n+1}(a,t) = (b_{i,k} a^i t^{k-i})_{0 \le i,k \le n},$$
$$Q_{n+1}(a,t) = (b_{(n-i),k} a^{n-i} t^{k-n+i})_{0 \le i,k \le n}$$

and for any elements $a \in \mathbb{F}_q^*$ and $t_1, t_2 \in \mathbb{F}_q$ with $t_1 \ne t_2$, define the matrix

$$R_{n+1}(a,t_1,t_2) = (c_{i,k} a^i)_{0 \le i,k \le n}.$$

THEOREM 5.2
The group $PO_{n+1}(\mathbb{F}_q)$ consists of the following $q(q^2 - 1)$ matrices:

$$T_{n+1}(a,t), \quad Q_{n+1}(a,t), \quad R_{n+1}(a,t_1,t_2)$$

where $a \in \mathbb{F}_q^*$, $t, t_1, t_2 \in \mathbb{F}_q$ with $t_1 \neq t_2$. The number of elements for these three types is $q(q-1)$, $q(q-1)$, $q(q-1)^2$, respectively.

REMARK 5.1 It can be proved by a group-theoretic approach that the group $PO_{n+1}(\mathbb{F}_q)$ is isomorphic to the group $PGL_2(\mathbb{F}_q)$ when $n > 0$ and $q \geq n$. ▯

PROOF Since

$$|PO_{n+1}(\mathbb{F}_q)| = \frac{|PGL_{n+1}(\mathbb{F}_q)|}{|\mathscr{E}|} = q(q^2 - 1),$$

it is only necessary to show that the set of points in Equation (5.2) is carried into itself by any matrix in $T_{n+1}(a,t)$, $Q_{n+1}(a,t)$, $R_{n+1}(a,t_1,t_2)$. The last conclusion of the theorem is obvious.

We have

$$\sum_{i=0}^{n} b_{i,k} a^i t^{k-i} \alpha^i = \sum_{i=0}^{k} C_k^i (a\alpha)^i t^{k-i} = (t + a\alpha)^k,$$

hence,

$$(1, \alpha, \cdots, \alpha^n) T_{n+1}(a,t) = (1, (t+a\alpha), \cdots, (t+a\alpha)^n).$$

We also have

$$(0, \cdots, 0, 1) T_{n+1}(a,t) = (0, \cdots, 0, a^n).$$

This shows that $T_{n+1}(a,t) \in PO_{n+1}(\mathbb{F}_q)$.

Similarly, we have

$$(1, \alpha, \cdots, \alpha^n) Q_{n+1}(a,t) = \begin{cases} a^n (1, t + a/\alpha, \cdots, (t + a/\alpha)^n), & \alpha \neq 0 \\ (0, \cdots, 0, a^n), & \alpha = 0 \end{cases}$$

and

$$(0, \cdots, 0, 1) Q_{n+1}(a,t) = (1, t, \cdots, t^n).$$

Therefore, $Q_{n+1}(a,t) \in PO_{n+1}(\mathbb{F}_q)$.

Finally, since

$$\sum_{i=0}^{n} c_{i,k} a^i \alpha^i = \sum_{i=0}^{n} \sum_{j=max(0,i-k)}^{min(i,n-k)} C_{n-k}^j C_k^{i-j} t_1^{k-(i-j)} t_2^{i-j} (a\alpha)^i$$

$$= \sum_{j=0}^{n-k} C_{n-k}^j (a\alpha)^j \sum_{s=0}^{k} C_k^s t_1^{k-s} (t_2 a\alpha)^s$$

$$= (1 + a\alpha)^{n-k} (t_1 + t_2 a\alpha)^k,$$

then

$$(1, \alpha, \cdots, \alpha^n) R_{n+1}(a, t_1, t_2)$$

$$= \begin{cases} (1 + a\alpha)^n (1, \frac{t_1 + t_2 a\alpha}{1 + a\alpha}, \cdots, \left(\frac{t_1 + t_2 a\alpha}{1 + a\alpha}\right)^n), & 1 + a\alpha \neq 0 \\ (0, \cdots, 0, (t_1 - t_2)^n), & 1 + a\alpha = 0 \end{cases}$$

Also

$$(0, \cdots, 0, 1) R_{n+1}(a, t_1, t_2) = a^n (1, t_2, \cdots, t_2^n).$$

Therefore, $R_{n+1}(a, t_1, t_2) \in PO_{n+1}(\mathbb{F}_q)$. Thus Theorem 5.2 is proved now. □

The elements with the form $T_{n+1}(a, t)$ form a subgroup of $PO_{n+1}(\mathbb{F}_q)$ with the order $q(q+1)$. We denote this subgroup by T_{n+1}.

Now we assume that $(q, n) = 1$.

THEOREM 5.3

All elements of $PGL_{n+1}(\mathbb{F}_q)$ whose last two rows have the form

$$\begin{matrix} 0 \cdots 0 \; 1 \; x_{n-1,v+1} \cdots x_{n-1,u-1} \; 0 \; x_{n-1,u+1} \cdots x_{n-1,n} \\ 0 \cdots \qquad\qquad\qquad\qquad 0 \quad 1 \;\; x_{n,u+1} \quad\cdots \quad x_{n,n} \end{matrix} \quad (v < u)$$

or

$$\begin{matrix} 0 \cdots 0 \; 0 \quad\cdots\quad\quad\quad 0 \; 1 \; x_{n-1,v+1} \cdots x_{n-1,n} \\ 0 \cdots 0 \; 1 \; x_{n,u+1} \cdots \qquad\qquad\qquad\qquad x_{n,n} \end{matrix} \quad (u < v)$$

form a representation system for cosets of T_{n+1} in $PGL_{n+1}(F_q)$.

PROOF Let $Y = (y_{i,j})$ be an element of $PGL_{n+1}(\mathbb{F}_q)$. The last two rows of the product $T_{n+1}(a, t) \cdot Y$ have the form

$$\begin{matrix} a^{n-1}(y_{n-1,0} + nty_{n,0}) & \cdots & a^{n-1}(y_{n-1,n} + nty_{n,n}) \\ a^n y_{n,0} & \cdots & a^n y_{n,n} \end{matrix}$$

Therefore, any two elements, given in the theorem, corresponding to different values of u belong to different cosets. Now suppose that Y and $T_{n+1}(a, t) \cdot Y$ are two elements given in the theorem, corresponding to the same value of u, then it deduces that $a = 1, t = 0$, noticing that $(n, q) = 1$. Hence, $T_{n+1}(a, t) = I$. This shows that any two elements given in the theorem belong to different cosets.

The number of the elements given in the theorem is

$$\frac{q^{n+1} - 1}{q - 1} \cdot \frac{q^n - 1}{q - 1} \cdot \prod_{i=2}^{n} (q^{n+1} - q^i) = q^{n(n+1)/2 - 1}(q+1) \prod_{i=3}^{n+1} (q^i - 1),$$

which is equal to

$$\frac{|PGL_{n+1}(Fq)|}{|T_{n+1}|} = \frac{q^{n(n+1)/2} \prod_{i=2}^{n+1} (q^i - 1)}{q(q-1)}.$$

This completes the proof of the theorem. □

5.3 Encoding Rules ($n = 2$, q Odd)

We assume $n = 2$ and q is odd in this section. Letting $T(a,t), Q(a,t), R(a, t_1, t_2)$ denote $T_3(a,t), Q_3(a,t), R_3(a,t_1,t_2)$, respectively, we have

$$T(a,t) = \begin{pmatrix} 1 & t & t^2 \\ 0 & a & 2at \\ 0 & 0 & a^2 \end{pmatrix}, \quad Q(a,t) = \begin{pmatrix} 0 & 0 & a^2 \\ 0 & a & 2at \\ 1 & t & t^2 \end{pmatrix},$$

$$R(a,t_1,t_2) = \begin{pmatrix} 1 & t_1 & t_1^2 \\ 2a & (t_1+t_2)a & 2at_1t_2 \\ a^2 & t_2a^2 & t_2^2a^2 \end{pmatrix}$$

The representation system for cosets of T_3 in $PGL_3(\mathbb{F}_q)$ given in Theorem 5.3 consists of the following six families:

(I)
$$\begin{pmatrix} x_{00} & x_{01} & x_{02} \\ 0 & 1 & 0 \\ 0 & 0 & 1 \end{pmatrix} \quad x_{00} \neq 0,$$

(II)
$$\begin{pmatrix} x_{00} & x_{01} & x_{02} \\ 1 & x_{11} & 0 \\ 0 & 0 & 1 \end{pmatrix} \quad x_{00}x_{11} - x_{01} \neq 0,$$

(III)
$$\begin{pmatrix} x_{00} & x_{01} & x_{02} \\ 0 & 0 & 1 \\ 0 & 1 & x_{22} \end{pmatrix} \quad x_{00} \neq 0,$$

(IV)
$$\begin{pmatrix} x_{00} & x_{01} & x_{02} \\ 1 & 0 & x_{12} \\ 0 & 1 & x_{22} \end{pmatrix} \quad x_{02} - x_{01}x_{22} - x_{00}x_{12} \neq 0,$$

(V)
$$\begin{pmatrix} x_{00} & x_{01} & x_{02} \\ 0 & 0 & 1 \\ 1 & x_{21} & x_{22} \end{pmatrix} \quad x_{00}x_{21} - x_{01} \neq 0,$$

(VI)
$$\begin{pmatrix} x_{00} & x_{01} & x_{02} \\ 0 & 1 & x_{12} \\ 1 & x_{21} & x_{22} \end{pmatrix} \quad x_{00}x_{22} + x_{01}x_{12} - x_{02} - x_{00}x_{12}x_{21} \neq 0.$$

The size of each family is given by

$$|I| = q^2(q-1), \quad |II| = q^3(q-1), \quad |III| = q^3(q-1),$$
$$|IV| = q^4(q-1), \quad |V| = q^4(q-1), \quad |VI| = q^5(q-1).$$

Since
$$|PO_3(\mathbb{F}_q)/T_3| = \frac{q(q^2-1)}{q(q-1)} = q+1$$
every coset of $PO_3(\mathbb{F}_q)$ in $PLG_3(\mathbb{F}_q)$ consists of $q+1$ cosets of T_3 in $PLG_3(\mathbb{F}_q)$ {we will omit the term "in $PLG_3(\mathbb{F}_q)$" in the following}. We partition $q^2(q+1)(q^3-1)$ representations given in Theorem 5.3 into $q^2(q^3-1)$ subsets, each subset consisting of $q+1$ cosets of T_3 that form a coset of $PO_3(\mathbb{F}_q)$.

(II) is partitioned into two subsets

(II$_1$)
$$\begin{pmatrix} 0 & x_{01} & x_{02} \\ 1 & x_{11} & 0 \\ 0 & 0 & 1 \end{pmatrix} \quad x_{01} \neq 0,$$

(II$_2$)
$$\begin{pmatrix} x_{00} & x_{01} & x_{02} \\ 1 & x_{11} & 0 \\ 0 & 0 & 1 \end{pmatrix} \quad x_{00} \neq 0, x_{00}x_{11} - x_{01} \neq 0,$$

here
$$|II_1| = q^2(q-1), \qquad |II_2| = q^2(q-1)^2.$$

(IV) is partitioned into four subsets

(IV$_1$)
$$\begin{pmatrix} x_{00} & -x_{00}^2 & x_{02} \\ 1 & 0 & 2x_{00}x_{22} \\ 0 & 1 & x_{22} \end{pmatrix} \quad x_{02} - x_{00}^2 x_{22} \neq 0,$$

(IV$_2$)
$$\begin{pmatrix} x_{00} & -x_{00}^2 & x_{02} \\ 1 & 0 & x_{12} \\ 0 & 1 & x_{22} \end{pmatrix} \quad x_{12} - 2x_{00}x_{22} \neq 0, x_{02} + x_{00}^2 x_{22} - x_{00}x_{12} \neq 0,$$

(IV$_3$)
$$\begin{pmatrix} x_{00} & -x_{00}^2 + s^2 & x_{02} \\ 1 & 0 & x_{12} \\ 0 & 1 & x_{22} \end{pmatrix} \quad s \neq 0, x_{02} + (x_{00}^2 - s^2)x_{22} - x_{00}x_{12} \neq 0,$$

(IV$_4$)
$$\begin{pmatrix} x_{00} & -x_{00}^2 + r & x_{02} \\ 1 & 0 & x_{12} \\ 0 & 1 & x_{22} \end{pmatrix} \quad x_{02} + (x_{00}^2 - r)x_{22} - x_{00}x_{12} \neq 0,$$

where $r \in \mathbb{F}_q^*$ is nonsquare element, and so

$$|IV_1| = q^2(q-1), |IV_2| = q^2(q-1)^2, |IV_3| = q^3(q-1)^2/2, |IV_4| = q^3(q-1)^2/2.$$

(V) is partitioned into three subsets

(V$_1$)
$$\begin{pmatrix} 0 & x_{01} & x_{02} \\ 0 & 0 & 1 \\ 1 & x_{21} & x_{22} \end{pmatrix} \quad x_{01} \neq 0,$$

(V_2)
$$\begin{pmatrix} -r^2 & x_{01} & x_{02} \\ 0 & 0 & 1 \\ 1 & x_{21} & x_{22} \end{pmatrix} \quad r \neq 0, r^2 x_{21} + x_{01} \neq 0,$$

(V_3)
$$\begin{pmatrix} x_{00} & x_{01} & x_{02} \\ 0 & 0 & 1 \\ 1 & x_{21} & x_{22} \end{pmatrix} \quad x_{00} x_{21} - x_{01} \neq 0,$$

where $-x_{00} \in \mathbb{F}_q^*$ is a nonsquare element. Hence,

$$|V_1| = q^3(q-1), \qquad |V_2| = q^3(q-1)^2/2, \qquad |V_3| = q^3(q-1)^2/2.$$

(VI) is partitioned into eight subsets

(VI_1)
$$\begin{pmatrix} 0 & 0 & x_{02} \\ 0 & 1 & x_{12} \\ 1 & x_{21} & x_{22} \end{pmatrix} \quad x_{02} \neq 0,$$

(VI_2)
$$\begin{pmatrix} 0 & x_{01} & x_{02} \\ 0 & 1 & x_{12} \\ 1 & x_{21} & x_{22} \end{pmatrix} \quad x_{01} \neq 0, x_{01} x_{12} - x_{02} \neq 0,$$

(VI_3)
$$\begin{pmatrix} -r^2 & -r^2 x_{21} - r & x_{02} \\ 0 & 1 & -2r x_{22} \\ 1 & x_{21} & x_{22} \end{pmatrix} \quad r \neq 0, x_{02} - r^2 x_{22} \neq 0,$$

(VI_4)
$$\begin{pmatrix} -r^2 & -r^2 x_{21} - r & r^2 x_{22} \\ 0 & 1 & x_{12} \\ 1 & x_{21} & x_{22} \end{pmatrix} \quad r \neq 0, x_{12} + 2r x_{22} \neq 0,$$

(VI_5)
$$\begin{pmatrix} -r^2 & -r^2 x_{21} - r & x_{02} \\ 0 & 1 & x_{12} \\ 1 & x_{21} & x_{22} \end{pmatrix} \quad \begin{array}{l} r \neq 0, x_{02} - r^2 x_{22} \neq 0, x_{12} + 2r x_{22} \neq 0, \\ r(2r x_{22} + x_{12}) + (x_{02} - r^2 x_{22}) \neq 0, \end{array}$$

(VI_6)
$$\begin{pmatrix} -r^2 & x_{01} & x_{02} \\ 0 & 1 & x_{12} \\ 1 & x_{21} & x_{22} \end{pmatrix} \quad \begin{array}{l} r \neq 0, \quad x_{01} \neq -r^2 x_{21} \pm r, \\ x_{01} x_{12} - x_{02} + r^2 x_{12} x_{21} - r^2 x_{22} \neq 0, \end{array}$$

(VI_7)
$$\begin{pmatrix} x_{00} & x_{01} & x_{02} \\ 0 & 1 & x_{12} \\ 1 & x_{21} & x_{22} \end{pmatrix} \quad x_{00} x_{22} + x_{01} x_{12} - x_{02} - x_{00} x_{12} x_{21} \neq 0,$$

where $-x_{00} \in \mathbb{F}_q^*$ is a nonsquare element, and there exists $r \in \mathbb{F}_q$ and $s \in \mathbb{F}_q^*$ such that $x_{00} = s^2 - r^2$, $x_{01} = x_{00}x_{21} - r$,

$$(\text{VI}_8) \quad \begin{pmatrix} x_{00} & x_{01} & x_{02} \\ 0 & 1 & x_{12} \\ 0 & x_{21} & x_{22} \end{pmatrix} \quad x_{00}x_{22} + x_{01}x_{12} - x_{02} - x_{00}x_{12}x_{21} \neq 0,$$

where $-x_{00} \in \mathbb{F}_q^*$ is a nonsquare element, and there does not exist $r \in \mathbb{F}_q$ and $s \in \mathbb{F}_q^*$ such that $x_{00} = s^2 - r^2$, $x_{01} = x_{00}x_{21} - r$, hence

$$\begin{aligned} &|\text{VI}_1| = q^3(q-1), & &|\text{VI}_2| = q^3(q-1)^2, & &|\text{VI}_3| = q^2(q-1)^2, \\ &|\text{VI}_4| = q^2(q-1)^2, & &|\text{VI}_5| = q^2(q-1)^2(q-2), & &|\text{VI}_6| = q^3(q-1)^2(q-2)/2, \\ &|\text{VI}_7| = q^3(q-1)^3/4, & &|\text{VI}_8| = q^3(q-1)^2(q+1)/4. \end{aligned}$$

The following lemma is needed to calculate $|\text{VI}_7|$.

LEMMA 5.4

Let $x \in \mathbb{F}_q^*$, n be the number of methods of representing x as a difference of two squares. Then

$$n = \begin{cases} (q-3)/4, & \text{if } q \equiv 3 \pmod 4 \\ (q-5)/4, & \text{if } q \equiv 1 \pmod 4, \ x \text{ is a square} \\ (q-1)/4, & \text{if } q \equiv 1 \pmod 4, \ x \text{ is a nonsquare} \end{cases}$$

PROOF   Define an χ-multiplication character on $\mathbb{F}_q^*$ as

$$\chi(a) = \begin{cases} 1, & a \text{ is a square} \\ -1, & a \text{ is a nonsquare} \\ 0, & a = 0 \end{cases}$$

It is easy to see that $\chi(ab) = \chi(a)\chi(b)$ for $\forall\, a, b \in \mathbb{F}_q^*$, and

$$\sum_{a \in \mathbb{F}_q^*} \chi(a) = 0$$

Suppose $a, b \in \mathbb{F}_q^*$ are square elements. There exists an $r_0 \in \mathbb{F}_q^*$ such that $a = br_0^2$, and hence,

$$\sum_{r \in \mathbb{F}_q^*} \chi(a + r^2) = \sum_{r \in \mathbb{F}_q^*} \chi(br_0^2 + r^2) = \sum_{r \in \mathbb{F}_q^*} \chi(r_0^2)\chi(b + (r/r_0)^2)$$

$$= \sum_{r \in \mathbb{F}_q^*} \chi(b + r^2).$$

In the same way, the above equality also holds when $a, b \in \mathbb{F}_q^*$ are nonsquare elements. Let

$$C_+ = \sum_{r \in \mathbb{F}_q^*} \chi(a + r^2), \quad \text{for } a \in \mathbb{F}_q^* \text{ is a square element}$$

$$C_- = \sum_{r \in \mathbb{F}_q^*} \chi(a + r^2), \quad \text{for } a \in \mathbb{F}_q^* \text{ is a nonsquare element}$$

Then

$$\frac{q-1}{2}(C_+ + C_-) = \sum_{a \in \mathbb{F}_q^*} \sum_{r \in \mathbb{F}_q^*} \chi(a + r^2) = \sum_{r \in \mathbb{F}_q^*} \sum_{a \in \mathbb{F}_q^*} \chi(a + r^2)$$

$$= \sum_{r \in \mathbb{F}_q^*} \left(\sum_{a \in \mathbb{F}_q^*} \chi(a) - 1 \right) = 1 - q$$

So, $C_+ + C_- = -2$. Similarly,

$$\frac{q-1}{2}(C_+ - C_-) = \sum_{a \in \mathbb{F}_q^*} \chi(a) \sum_{r \in \mathbb{F}_q^*} \chi(a + r^2) = \sum_{a, r \in \mathbb{F}_q^*} \chi(a^2 + ar^2)$$

$$= \sum_{a, r \in \mathbb{F}_q^*} \chi(a + (a/r)^2) = \sum_{b \in \mathbb{F}_q^*} \sum_{a \in \mathbb{F}_q^*} \chi(a + b^2) = 1 - q$$

So, $C_+ - C_- = -2$. Therefore $C_+ = -2, C_- = 0$.

Suppose g is a generator of $\mathbb{F}_q^*$, that is, $-1 = g^{(q-1)/2}$. Then -1 is a nonsquare element when $q \equiv 3 \pmod 4$, while -1 is a square element when $q \equiv 1 \pmod 4$.

Consider $q \equiv 3 \pmod 4$. Suppose x is a square element. Then $-x$ is a nonsquare element. Hence,

$$C_+ = \sum_{r \in \mathbb{F}_q^*} \chi(x + r^2) = 2n - (q - 1 - 2n) = 4n - q + 1$$

and so we have $n = (q - 3)/4$. Suppose x is a nonsquare element. Then $-x$ is a square element. There are two values for r such that $x + r^2 = 0$. Hence,

$$C_- = \sum_{r \in \mathbb{F}_q^*} \chi(x + r^2) = 2n - (q - 3 - 2n) = 4n - q + 3$$

and so we also have $n = (q - 3)/4$.

Consider $q \equiv 1 \pmod 4$. Suppose x is a square element. Then $-x$ is also a square element. Hence,

$$C_+ = \sum_{r \in \mathbb{F}_q^*} \chi(x + r^2) = 2n - (q - 3 - 2n) = 4n - q + 3$$

so, we have $n = (q - 5)/4$. Suppose x is a nonsquare element. Then $-x$ is also a nonsquare element. Hence,

$$C_- = \sum_{r \in \mathbb{F}_q^*} \chi(x + r^2) = 2n - (q - 1 - 2n) = 4n - q + 1$$

so we have $n = (q - 1)/4$.   ∎

Using Lemma 5.4, $|VI_7|$ is calculated as follows. Suppose $-x_{00} \in \mathbb{F}_q^*$ is a nonsquare element. First, consider $q \equiv 3 \,(mod\,4)$. Then x_{00} is a square element, hence, it has $(q - 1)/2$ choices. For each fixed x_{00}, there exists $s \in \mathbb{F}_q^*$ such that $x_{00} = s^2$, in this case $r = 0$. From Lemma 5.4, there are $(q - 3)/4$ pairs of $(s, r) \in \mathbb{F}_q^* \times \mathbb{F}_q^*$ such that $x_{00} = s^2 - r^2$. So, there are $(q - 3)/2$ choices for $r \in \mathbb{F}_q^*$. For any fixed x_{21} (it has q choices) there are $1 + (q - 3)/2 = (q - 1)/2$ choices for x_{01}. Therefore, $|VI_7| = q^3(q - 1)^3/4$. The same result can be obtained when $q \equiv 1 \,(mod\,4)$.
As

$$|VI_7| + |VI_8| = q^4(q - 1)^2/2$$

$|VI_8|$ is also obtained.

LEMMA 5.5
For every element in (I), *there exist q elements in* (VI_1), *such that these $q + 1$ elements are in the same coset of $PO_3(\mathbb{F}_q)$.*

PROOF   The product of $\lambda \cdot R(a, t_1, t_2)$ with an element in (VI) (replace x_{ij} by y_{ij}) is

$$\lambda \cdot R(a, t_1, t_2) \begin{pmatrix} y_{00} & y_{01} & y_{02} \\ 0 & 1 & y_{12} \\ 1 & y_{21} & y_{22} \end{pmatrix} =$$

$$\lambda \begin{pmatrix} y_{00} + t_1^2 & y_{01} + t_1 + t_1^2 y_{21} & y_{02} + t_1 y_{12} + t_1^2 y_{22} \\ 2a(y_{00} + t_1 t_2) & a(2y_{01} + t_1 + t_2 + 2t_1 t_2 y_{21}) & a(2y_{02} + (t_1 + t_2)y_{12} + 2t_1 t_2 y_{22}) \\ a^2(y_{00} + t_2^2) & a^2(y_{01} + t_2 + t_2^2 y_{21}) & a^2(y_{02} + t_2 y_{12} + t^2 y_{22}) \end{pmatrix}$$

$$(5.5)$$

where $y_{00}y_{22} + y_{01}y_{12} - y_{02} - y_{00}y_{12}y_{21} \neq 0$. Assume $y_{00} = y_{01} = 0$, then $y_{02} \neq 0$. Let (VI_{11}) be the subset of (VI_1) consisting of elements with $y_{12} \neq 0$, and let (VI_{12}) be the subset of (VI_1) consisting of elements with $y_{12} = 0$. When $y_{12} \neq 0$, let

$$t_1 = -\frac{2y_{02}}{y_{12}}, \; t_2 = 0, \; a = -\frac{2}{y_{12}}, \; \lambda = \frac{y_{12}^2}{4y_{02}}$$

the right side of Equation (5.5) is

$$\begin{pmatrix} y_{02} & y_{02}y_{21} - y_{12}/2 & y_{02}y_{22} - y_{12}^2/4 \\ 0 & 1 & 0 \\ 0 & 0 & 1 \end{pmatrix}$$

This is an element of (I). Further, assume

$$y_{02} = x_{00} \, (\neq 0), \quad y_{02}y_{21} - y_{12}/2 = x_{01}, \quad y_{02}y_{22} - y_{12}^2 = x_{02}$$

then we have

$$y_{02} = x_{00},$$
$$y_{21} = (x_{01} + y_{12}/2)/x_{00},$$
$$y_{22} = (x_{02} + y_{12}^2/4)/x_{00}.$$

Here y_{12} is an arbitrary element of $\mathbb{F}_q^*$. Therefore, every element of (I) and $q-1$ element of (VI_{11}) are in the same coset of $PO_3(\mathbb{F}_q^*)$.

When $y_{12} = 0$, since

$$y_{02}^{-1} Q(y_{02}, 0) \begin{pmatrix} 0 & 0 & y_{02} \\ 0 & 1 & 0 \\ 1 & y_{21} & y_{22} \end{pmatrix} = \begin{pmatrix} y_{02} & y_{02}y_{21} & y_{02}y_{22} \\ 0 & 1 & 0 \\ 0 & 0 & 1 \end{pmatrix}$$

then every element in (I) is in the same coset with one element of (VI_{12}). □

LEMMA 5.6

For every element in (II_1), there exist one element of (IV_1) and $q-1$ elements of (VI_4), such that these $q+1$ elements are in the same coset of $PO_3(\mathbb{F}_q)$.

PROOF   The product of the element $\lambda \cdot Q(a,t)$ with an element of (IV) is

$$\lambda \cdot Q(a,t) \begin{pmatrix} y_{00} & y_{01} & y_{02} \\ 1 & 0 & y_{12} \\ 0 & 1 & y_{22} \end{pmatrix} = \lambda \begin{pmatrix} 0 & a^2 & a^2 y_{22} \\ a & 2at & a(y_{12} + 2ty_{22}) \\ y_{00} + t & y_{01} + t^2 & y_{02} + ty_{12} + t^2 y_{22} \end{pmatrix} \quad (5.6)$$

where $y_{02} - y_{00}y_{12} - y_{01}y_{22} \neq 0$. Assume $y_{01} = -y_{00}^2, y_{12} = 2y_{00}y_{22}$. Let

$$t = -y_{00}, \quad a = y_{02} - y_{00}^2 y_{22}, \quad \lambda = (y_{02} - y_{00}^2 y_{22})^{-1}$$

the right side of Equation (5.6) becomes

$$\begin{pmatrix} 0 & y_{02} - y_{00}^2 y_{22} & y_{22}(y_{02} - y_{00}^2 y_{22}) \\ 1 & -2y_{00} & 0 \\ 0 & 0 & 1 \end{pmatrix}$$

which is in (II_1). Since the following system

$$y_{02} - y_{00}^2 y_{22} = x_{01}, \, (\neq 0)$$
$$y_{22}(y_{02} - y_{00}^2 y_{22}) = x_{02},$$
$$-2y_{00} = x_{11}$$

determines uniquely y_{00}, y_{02}, y_{22}, so every element in (II_1) is in the same coset of $PO_3(\mathbb{F}_q)$ with an element of (IV_1). Assume that in Equation (5.5)

$y_{00} = -r^2 (\neq 0), y_{01} = -r^2 y_{21} - r, y_{02} = r^2 y_{22}$ and $y_{12} + 2ry_{22} \neq 0$, i.e., assume the element in the left side of Equation (5.5) belongs to (VI$_4$). Let

$$t_1 = -r, \qquad\qquad t_2 = r,$$

$$a = -\frac{4r}{y_{12} + 2ry_{22}}, \qquad \lambda = \frac{y_{12} + 2ry_{22}}{16r^3}.$$

The right side of Equation (5.5) is

$$\begin{pmatrix} 0 & -\frac{y_{12}+2ry_{22}}{8r} & \frac{(2ry_{22}+y_{12})(2ry_{22}-y_{12})}{16r^2} \\ 1 & y_{21} + \frac{1}{2r} & 0 \\ 0 & 0 & 1 \end{pmatrix}$$

which is in (II$_1$). From the following system

$$-\frac{2ry_{22} + y_{12}}{8r} = x_{01}, \; (\neq 0)$$

$$\frac{(2ry_{22} + y_{12})(2ry_{22} - y_{12})}{16r^2} = x_{02},$$

$$y_{21} + \frac{1}{2r} = x_{12},$$

we have

$$y_{21} = x_{12} - \frac{1}{2r},$$

$$y_{12} = -4r^2 x_{01} + \frac{x_{02}}{x_{01}},$$

$$y_{22} = -2rx_{01} - \frac{x_{02}}{2rx_{01}}.$$

Here, $r \in \mathbb{F}_q^*$ is an arbitrary element. So every element in (II$_1$) is in the same coset of $PO_3(\mathbb{F}_q)$ with $q - 1$ elements of (VI$_4$). $\quad\square$

LEMMA 5.7
For every element in (II$_2$), *there exist one element in* (IV$_2$) *and* $q-2$ *elements in* (VI$_5$), *such that these* $q + 1$ *elements are in the same coset of* $PO_3(\mathbb{F}_q)$.

PROOF   The product of the element $\lambda \cdot R(a, t_1, t_2)$ with an element in (IV) is

$$\lambda \cdot R(a, t_1, t_2) \begin{pmatrix} y_{00} & y_{01} & y_{02} \\ 1 & 0 & y_{12} \\ 0 & 1 & y_{22} \end{pmatrix} =$$

$$\lambda \begin{pmatrix} y_{00} + t_1 & y_{01} + t_1^2 & y_{02} + t_1 y_{12} + t_1^2 y_{22} \\ a(2y_{00} + t_1 + t_2) & 2a(y_{01} + t_1 t_2) & a(2y_{02} + (t_1+t_2)y_{12} + 2t_1 t_2 y_{22}) \\ a^2(y_{00} + t_2) & a^2(y_{01} + t_2^2) & a^2(y_{02} + t_2 y_{12} + t_2^2 y_{22}) \end{pmatrix} \qquad (5.7)$$

where $y_{02} - y_{00}y_{12} - y_{01}y_{22} \neq 0$. Suppose $y_{01} = -y_{00}^2$, $y_{12} - 2y_{00}y_{22} \neq 0$, i.e., the element of (IV) on the left of Equation (5.7) is in (IV$_2$). Let

$$t_1 = \frac{2y_{02} - y_{00}y_{12}}{2y_{00}y_{22} - y_{12}},$$

$$t_2 = -y_{00},$$

$$a = \frac{2}{2y_{00}y_{22} - y_{12}},$$

$$\lambda = \frac{(2y_{00}y_{22} - y_{12})^2}{4(y_{02} - y_{00}y_{12} - y_{01}y_{22})}.$$

The matrix in the right-hand side of Equation (5.7) is rewritten as

$$\begin{pmatrix} y_{00}y_{22} - y_{12}/2 & y_{02} - y_{00}^2 y_{22} & y_{02}y_{22} - y_{12}^2/4 \\ 1 & -2y_{00} & 0 \\ 0 & 0 & 1 \end{pmatrix}$$

which is in (II$_2$). Let

$$-2y_{00} = x_{11},$$

$$y_{00}y_{22} - y_{12}/2 = x_{00},$$

$$y_{02} - y_{00}^2 y_{22} = x_{01},$$

$$y_{02}y_{22} - y_{12}^2/4 = x_{02},$$

and $x_{00}x_{11} - x_{01} \neq 0$. Then the following is obtained from the first three equations:

$$y_{00} = -x_{11}/2,$$

$$y_{12} = -2x_{00} - x_{11}y_{22},$$

$$y_{02} = x_{01} + x_{11}^2 y_{22}/4.$$

Replacing them into the 4th equation, we have

$$y_{22}(x_{01} + x_{11}^2 y_{22}/4) - (2x_{00} + x_{11}y_{22})^2/4 = x_{02},$$

hence,

$$y_{22} = \frac{x_{00}^2 + x_{02}}{x_{01} - x_{00}x_{11}}.$$

So every element in (II$_2$) is in the same coset of $PO_3(\mathbb{F}_q)$ with one element in (IV$_2$).

Next, the product of element $\lambda \cdot Q(a,t)$ and an element of (VI) is

$$\lambda \cdot Q(a,t) \begin{pmatrix} y_{00} & y_{01} & y_{02} \\ 0 & 1 & y_{12} \\ 1 & y_{21} & y_{22} \end{pmatrix} = \lambda \begin{pmatrix} a^2 & a^2 y_{21} & a^2 y_{22} \\ 2at & a(1 + 2ty_{21}) & a(y_{12} + 2ty_{22}) \\ y_{00} + t^2 & y_{01} + t + t^2 y_{21} & y_{12} + ty_{12} + t^2 y_{22} \end{pmatrix} \quad (5.8)$$

where $y_{00}y_{22} + y_{01}y_{12} - y_{02} - y_{00}y_{12}y_{21} \neq 0$. Suppose $y_{00} = -r^2 \neq 0, y_{01} = -r^2 y_{21} - r$, $y_{21} = -2ry_{22}$, and $y_{02} \neq r^2 y_{22}$, i.e., the element of (VI) on the left-hand side of Equation (5.8) is in (VI$_3$). Let

$$t = r, \quad a = \frac{y_{02} - r^2 y_{22}}{2r}, \quad \lambda = \frac{1}{y_{02} - r^2 y_{22}}.$$

Then the matrix on the right-hand side of Equation (5.8) is rewritten as

$$\begin{pmatrix} \frac{y_{02} - r^2 y_{22}}{4r^2} & \frac{y_{21}(y_{02} - r^2 y_{22})}{4r^2} & \frac{y_{22}(y_{02} - r^2 y_{22})}{4r^2} \\ 1 & y_{21} + \frac{1}{2r} & 0 \\ 0 & 0 & 1 \end{pmatrix}$$

which is in (II$_2$). Let

$$\frac{y_{02} - r^2 y_{22}}{4r^2} = x_{00}, \; (\neq 0),$$

$$\frac{y_{21}(y_{02} - r^2 y_{22})}{4r^2} = x_{01},$$

$$\frac{y_{22}(y_{02} - r^2 y_{22})}{4r^2} = x_{02},$$

$$y_{21} + \frac{1}{2r} = x_{11}$$

and $x_{00}x_{11} - x_{01} \neq 0$. Then,

$$y_{21} = \frac{x_{01}}{x_{00}}, \quad r = \frac{x_{00}}{2(x_{00}x_{11} - x_{01})},$$
$$y_{22} = \frac{x_{02}}{x_{00}}, \quad y_{02} = 4r^2 x_{00} + \frac{r^2 x_{02}}{x_{00}}.$$

Hence, every element in (II$_2$) is in the same coset of $PO_3(\mathbb{F}_q)$ with one element in (VI$_3$).

In Equation (5.5), let $y_{00} = -r^2 \neq 0, y_{01} = -r^2 y_{21} - r, y_{02} \neq r^2 y_{22}, y_{12} \neq -2ry_{22}$, i.e., the element of (VI) on the left-hand side of Equation (5.5) is in (VI$_5$). Let

$$t_1 = -\frac{2y_{02} + ry_{12}}{y_{12} + 2ry_{22}}, \quad a = -\frac{4r}{y_{12} + 2ry_{22}},$$
$$t_2 = r, \quad \lambda = \frac{(y_{12} + 2ry_{22})^2}{16r^2(y_{02} + ry_{12} + r^2 y_{22})}.$$

Note that $t_1 \neq t_2$, the matrix in the right-hand side of Equation (5.5) is rewritten as

$$\begin{pmatrix} \frac{y_{02} - r^2 y_{22}}{4r^2} & -\frac{y_{12} + 2ry_{22}}{8r^2} + \frac{y_{21}(y_{02} - r^2 y_{22})}{4r^2} & \frac{4y_{02}y_{22} - y_{12}^2}{16r^2} \\ 1 & y_{21} + \frac{1}{2r} & 0 \\ 0 & 0 & 1 \end{pmatrix}$$

which is in (II$_2$). Suppose

$$x_{11} = y_{21} + \frac{1}{2r},$$
$$x_{00} = \frac{y_{02} - r^2 y_{22}}{4r^2},$$
$$x_{01} = -\frac{y_{12} + 2ry_{22}}{8r^2} + \frac{y_{21}(y_{02} - r^2 y_{22})}{4r^2},$$
$$x_{02} = \frac{4y_{02}y_{22} - y_{12}^2}{16r^2}$$

and $x_{00} \neq 0, x_{00}x_{11} - x_{01} \neq 0$. The following is obtained from the first three equations

$$y_{21} = x_{11} - \frac{1}{2r},$$
$$y_{02} = 4r^2 x_{00} + r^2 y_{22},$$
$$y_{12} = -2ry_{22} + 4r\left[2r(x_{00}x_{11} - x_{01}) - x_{00}\right].$$

Replacing them into the fourth equation, we have

$$y_{22} = \frac{x_{02} - \left[2r(x_{00}x_{11} - x_{01}) - x_{00}\right]^2}{2r(x_{00}x_{01} - x_{01})},$$

since $rx_{00} \neq 0$, so $y_{02} \neq r^2 y_{22}$. Further, if

$$r \neq \frac{x_{00}}{2(x_{00}x_{11} - x_{01})}$$

then $y_{12} \neq -2ry_{22}$. Therefore, there are $q - 2$ possible values for r. By now we have proved that every element in (II$_2$) is in the same coset of $PO_3(\mathbb{F}_q)$ with $q - 2$ elements in (VI$_5$). The Lemma 5.7 is completely proved. $\square$

LEMMA 5.8

For every element in (III), *there exist* $q-1$ *elements in* (VI$_2$) *and one element in* (V), *such that these* $q + 1$ *elements are in the same coset of* $PO_3(\mathbb{F}_q)$.

PROOF In Equation (5.5), let $y_{00} = 0$, $y_{01} \neq 0$, hence, the element of (VI) on the left-hand side of Equation (5.5) is in (VI$_2$). Let

$$t_1 = -2y_{01}, \quad t_2 = 0, \quad a = \frac{2(y_{02} - y_{01}y_{12})}{y_{01}}, \quad \lambda = \frac{y_{01}}{4(y_{02} - y_{01}y_{12})^2}.$$

The matrix on the right-hand side of Equation (5.5) is rewritten as

$$\begin{pmatrix} \frac{y_{01}}{(y_{12}-y_{02}/y_{01})^2} & \frac{y_{01}y_{21}-1/4}{(y_{12}-y_{02}/y_{01})^2} & \frac{y_{01}y_{22}-y_{12}/2+y_{02}/(4y_{01})}{(y_{12}-y_{02}/y_{01})^2} \\ 0 & 0 & 1 \\ 0 & 1 & y_{02}/y_{01} \end{pmatrix}$$

which is in (III). Suppose

$$x_{00} = \frac{y_{01}}{(y_{12}-y_{02}/y_{01})^2},$$
$$x_{01} = \frac{y_{01}y_{21}-1/4}{(y_{12}-y_{02}/y_{01})^2},$$
$$x_{02} = \frac{y_{01}y_{22}-y_{12}/2+y_{02}/(4y_{01})}{(y_{12}-y_{02}/y_{01})^2},$$
$$x_{22} = \frac{y_{02}}{y_{01}}.$$

Then it is obtained that

$$y_{01} = x_{00}(y_{12} - x_{22})^2,$$
$$y_{21} = \frac{x_{01}(y_{12}-x_{22})^2+1/4}{x_{00}(y_{12}-x_{22})^2},$$
$$y_{22} = \frac{x_{02}(y_{12}-x_{22})^2+y_{12}/2-x_{22}/4}{x_{00}(y_{12}-x_{22})^2}.$$

Here, y_{12} is chosen such that $y_{12} \neq x_{22}$. So every element in (III) is in the same coset of $PO_3(\mathbb{F}_q)$ with $q-1$ elements in (VI$_2$). Since

$$y_{01}Q(y_{01},0)\begin{pmatrix} 0 & y_{01} & y_{01} \\ 0 & 0 & 1 \\ 1 & y_{21} & y_{22} \end{pmatrix} = \begin{pmatrix} y_{01} & y_{01}y_{21} & y_{01}y_{22} \\ 0 & 0 & 1 \\ 0 & 1 & y_{02}/y_{01} \end{pmatrix},$$

then every element in (III) is in the same coset of $PO_3(\mathbb{F}_q)$ with one element in (V$_1$).   □

LEMMA 5.9

For every element in (IV$_3$), *there exist another element in* (IV$_3$) *and* $q-3$ *elements in* (VI$_6$), *such that these* $q-1$ *elements are in the same coset of* $PO_3(\mathbb{F}_q)$. *For every element in* (IV$_4$), *there exist another element in* (IV$_4$) *and* $q-1$ *elements in* (VI$_6$), *such that these* $q+1$ *elements are in the same coset of* $PO_3(\mathbb{F}_q)$.

PROOF   In Equation (5.5), let $y_{00} = -r^2 \neq 0$ and

$$y_{01} \neq -r^2 y_{21} \pm r. \tag{5.9}$$

Then the element on the left-hand side of Equation (5.5) is in (VI$_6$). Consider the following two cases.

(1) If $y_{21} \neq \pm 1/2r$, take $t_2^{(1)} = r, t_2^{(2)} = -r$,

$$t_1^{(i)} = -\frac{2y_{01} + t_2^{(i)}}{2t_2^{(i)}y_{21} + 1},$$

$$a^{(i)} = -\frac{4t_2^{(i)}}{2t_2^{(i)}y_{21} + 1},$$

$$\lambda^{(i)} = \frac{(2t_2^{(i)}y_{21} + 1)^2}{16r^2(t_2^{(i)} + y_{01} + r^2 y_{21})}$$

for $i = 1, 2$. Then the matrix in the right-hand side of Equation (5.5) is rewritten as

$$\begin{pmatrix} \frac{y_{01}}{4r^2} - \frac{y_{21}}{4} & \frac{4y_{01}y_{21}-1}{16r^2} & \star \\ 1 & 0 & \star \\ 0 & 1 & \star \end{pmatrix} \tag{5.10}$$

(2) If $y_{21} = \pm 1/2r$, we may assume that $y_{21} = 1/2r$ (replace r by $-r$ if $y_{21} = -1/2r$), $t_2 = r$, and t_1, a λ are defined as above. Now the matrix in the right-hand side of Equation (5.5) is rewritten as

$$\begin{pmatrix} \frac{2y_{01}-r}{8r^2} & \frac{2y_{01}-r}{16r^3} & \star \\ 1 & 0 & \star \\ 0 & 1 & \star \end{pmatrix}. \tag{5.11}$$

Note that we have another equality in this case:

$$\frac{2}{2y_{01}-r}\cdot Q\left(\frac{r-2y_{01}}{4r},-r\right)\begin{pmatrix}-r^2 & y_{01} & y_{02}\\0 & 1 & y_{12}\\1 & 1/2r & y_{22}\end{pmatrix}=\begin{pmatrix}\frac{2y_{01}-r}{8r^2} & \frac{2y_{01}-r}{16r^3} & \star\\1 & 0 & \star\\0 & 1 & \star\end{pmatrix}\quad(5.12)$$

Given an element of (IV_3) or (IV_4),

$$\begin{pmatrix}x_{00} & -x_{00}^2+k\ x_{02}\\1 & 0 & x_{12}\\0 & 1 & x_{22}\end{pmatrix}\quad(5.13)$$

here $k\neq 0$, if the matrix in Equation (5.10) equals the matrix in Equation (5.13) then

$$\frac{y_{01}}{4r^2}-\frac{y_{21}}{4}=x_{00},\qquad -\frac{y_{01}}{4r^2}\cdot\frac{y_{21}}{4}=\frac{x_{00}^2-k}{4}-\frac{1}{64r^2}.\quad(5.14)$$

Since

$$\left(\frac{y_{01}}{4r^2}+\frac{y_{21}}{4}\right)^2=k+\frac{1}{16r^2},$$

the system of Equations (5.14) has two solutions if and only if $k+1/(16r^2)$ is a nonzero square in $\mathbb{F}_q$, and has one solution if and only if $k+1/(16r^2)=0$. Let (y_{01}^*,y_{21}^*) be a solution of Equation (5.14). From

$$\left(\frac{y_{01}^*}{4r^2}+\frac{y_{21}^*}{4}\right)^2\neq\frac{1}{16r^2},$$

This shows that (y_{01}^*,y_{21}^*) satisfies the condition in Equation (5.9).
In the case of $y_{21}^*\neq\pm1/2r$, let

$$\begin{pmatrix}y_{02}^{*\,(i)}\\y_{12}^{*\,(i)}\\y_{22}^{*\,(i)}\end{pmatrix}=\left(\lambda^{(i)}\cdot R\left(a^{(i)},t_1^{(i)},t_2^{(i)}\right)\right)^{-1}\begin{pmatrix}x_{02}\\x_{12}\\x_{22}\end{pmatrix}$$

for $i=1,2$. This shows that there are two elements in (VI_6) which are in the same coset of $PO_3(\mathbb{F}_q)$ with the element in Equation (5.13).
In the case of $y_{12}^*=\pm1/2r$, we can find only one element in (VI_6), which is in the same coset of $PO_3(\mathbb{F}_q)$ with the element in Equation (5.13), by the above way. But in Equation (5.12), letting

$$\begin{pmatrix}y_{02}\\y_{12}\\y_{22}\end{pmatrix}=\frac{2y_{01}^*-r}{2}\cdot Q\left(\frac{r-2y_{01}^*}{4r},-r\right)^{-1}\begin{pmatrix}x_{02}\\x_{12}\\x_{22}\end{pmatrix},$$

we can find another element of (VI_6) which is in the same coset of $PO_3(\mathbb{F}_q)$ with the element in Equation (5.13). So we have proved that for any solution to Equation (5.14) and any element of (IV$_3$) or (IV$_4$) in Equation (5.13), in whose coset of $PO_3(\mathbb{F}_q)$ we can always find two elements of (VI$_6$).

Assuming that the element in Equation (5.13) is in (IV$_3$), then $k \neq 0$ is a square. When $q \equiv 1 \pmod 4$, -1 is also a square. There is a unique $v_0 \in \mathbb{F}_q$ such that $k = -v_0^2$. Letting $r^2 = 1/(16v_0^2)$, the system of Equations (5.14) has one solution in this case. From Lemma 5.4, there exist $(q-5)/4$ pairs (u^2, v^2), $u \neq 0, v \neq 0$, satisfying $k = u^2 - v^2$. Let $r^2 = 1/(16v^2)$. For each pair (u^2, v^2), there exist two solutions to Equation (5.14). Hence, there are totally

$$1 + \frac{q-5}{2} = \frac{q-3}{2}$$

solutions to Equation (5.14). When $q \equiv 3 \pmod 4$, there is no $v \neq 0$ satisfying $k = -v^2$, but there are $(q-3)/4$ pairs of (u^2, v^2), $u \neq 0, v \neq 0$ satisfying $k = u^2 - v^2$. Therefore, there are $(q-3)/2$ solutions to Equation (5.14). From the above discussion we know that there are $q-3$ elements in (VI$_6$) which are in the same coset of $PO_3(\mathbb{F}_q)$ with the element in Equation (5.13).

Assuming that the element in Equation (5.13) is in (IV$_4$), then $k \neq 0$ is nonsquare. Using Lemma 5.4 and the similar discussion, one can also prove that there are $q-1$ elements in (VI$_6$) that are in the same coset of $PO_3(\mathbb{F}_q)$ with the element in Equation (5.13).

In Equation (5.13), if $x_{00} = 0$, therefore, $x_{02} \neq kx_{22}$, then

$$k^{-1} \cdot Q(k,0) \begin{pmatrix} 0 & k & x_{02} \\ 1 & 0 & x_{12} \\ 0 & 1 & x_{22} \end{pmatrix} = \begin{pmatrix} 0 & k & kx_{22} \\ 1 & 0 & x_{12} \\ 0 & 1 & x_{02}/k \end{pmatrix}.$$

The matrix on the right-hand side is an element in (IV$_3$) {or (IV$_4$)}, which is different from the matrix in (IV$_3$) {or (IV$_4$)}, on the left-hand side.

In Equation (5.13), if $x_{00} \neq 0$, we have

$$\frac{x_{00}^2}{k} R\left(x_{00}^{-1}, \frac{k}{x_{00}} - x_{00}, -x_{00}\right) = \begin{pmatrix} x_{00}^2/k & x_{00}(1 - x_{00}^2/k) & k(1 - x_{00}^2/k)^2 \\ 2x_{00}/k & 1 - 2x_{00}^2/k & 2x_{00}(x_{00}^2/k - 1) \\ 1/k & -x_{00}/k & x_{00}^2/k \end{pmatrix} \quad (5.15)$$

Note that $(x_{00}^2/k)^2 R(x_{00}^{-1}, \frac{k}{x_{00}} - x_{00}, -x_{00})^2 = I$. (When $x_{00} = 0$, the matrix on the right-hand side is just $k^{-1}Q[k, 0]$.) Therefore,

$$\frac{x_{00}^2}{k} R\left(\frac{1}{x_{00}}, \frac{k}{x_{00}} - x_{00}, -x_{00}\right) \begin{pmatrix} x_{00} & -x_{00}^2 + k & x_{02} \\ 1 & 0 & x_{12} \\ 0 & 1 & x_{22} \end{pmatrix} = \begin{pmatrix} x_{00} & -x_{00}^2 + k & x_{02}' \\ 1 & 0 & x_{12}' \\ 0 & 1 & x_{22}' \end{pmatrix}.$$

The matrix on the right-hand side is in (IV$_3$) {or (IV$_4$)}, and it is different from the matrix in (IV$_3$) {or (IV$_4$)}, since Equation (5.15) $\neq I$. So, there are two elements of (IV$_3$), and of (IV$_4$) as well, in the same coset. ∎

LEMMA 5.10

For every element of (V_2), *there exist another element of* (V_2), *two elements of* (IV_3), *and* $q - 3$ *elements of* (VI_6), *such that these* $q + 1$ *elements are in the same coset of* $PO_3(\mathbb{F}_q)$.

PROOF By Lemma 5.9, we only need to show that for every element of (V_2), there exist another element of (V_2) and two elements of (IV_3), such that they are in the same coset.

When $x_{00} \neq 0$, since

$$
x_{00}^{-1} \cdot Q(x_{00}, 0) \begin{pmatrix} x_{00} & x_{01} & x_{02} \\ 0 & 0 & 1 \\ 1 & x_{21} & x_{22} \end{pmatrix} = \begin{pmatrix} x_{00} & x_{00}x_{21} & x_{00}x_{22} \\ 0 & 0 & 1 \\ 1 & x_{01}/x_{00} & x_{02}/x_{00} \end{pmatrix}
$$

and $x_{01} \neq x_{00}x_{21}$, so every element in (V_2) {or (V_3)}, is in the same coset of $PO_3(\mathbb{F}_q)$ as another element of (V_2) {or (V_3)}.

In Equation (5.7), let $y_{01} = -y_{00}^2 + s^2$, $s \neq 0$, i.e., the elements of (IV) on the left-hand side of Equation (5.7) is in (IV_3), let

$$
t_1 = s - y_{00}, \quad t_2 = -s - y_{00},
$$
$$
a = -\frac{2(y_{02} - y_{00}y_{12} - y_{01}y_{22})}{s},
$$
$$
\lambda = -\frac{s}{4(y_{02} - y_{00}y_{12} - y_{01}y_{22})^2}.
$$

The right-hand side of Equation (5.7) is rewritten as

$$
\begin{pmatrix} \frac{-s^2}{4(y_{02}-y_{00}y_{12}-y_{01}y_{22})^2} & \frac{s^2(y_{00}-s)}{2(y_{02}-y_{00}y_{12}-y_{01}y_{22})^2} & \frac{-s(y_{02}-y_{00}y_{12}-y_{01}y_{22})-s^2\left(y_{12}+2(s-y_{00})y_{22}\right)}{4(y_{02}-y_{00}y_{12}-y_{01}y_{22})^2} \\ 0 & 0 & 1 \\ 1 & -2(s+y_{00}) & y_{12} - 2(s+y_{00})y_{22} - \frac{y_{02}-y_{00}y_{12}-y_{01}y_{22}}{s} \end{pmatrix}
$$
$$
\tag{5.16}
$$

which is in (V_2).

For a given element in (V_2),

$$
\begin{pmatrix} -r^2 & x_{01} & x_{02} \\ 0 & 0 & 1 \\ 1 & x_{21} & x_{22} \end{pmatrix}
\tag{5.17}
$$

where $r \neq 0, x_{01} \neq -r^2 x_{21}$. If Equations (5.16) and (5.17) are the same, then

$$r^2 = \frac{s^2}{4(y_{02} - y_{00}y_{12} - y_{01}y_{22})^2},$$

$$x_{01} = \frac{s^2(y_{00} - s)}{2(y_{02} - y_{00}y_{12} - y_{01}y_{22})^2},$$

$$x_{02} = \frac{-s}{4(y_{02} - y_{00}y_{12} - y_{01}y_{22})} - \frac{s^2(y_{12} + 2(s - y_{00})y_{22})}{4(y_{02} - y_{00}y_{12} - y_{01}y_{22})^2},$$

$$x_{22} = y_{12} - 2(s + y_{00})y_{22} - \frac{y_{02} - y_{00}y_{12} - y_{01}y_{22}}{s},$$

$$x_{21} = -2(s + y_{00}).$$

Hence,

$$\frac{s}{2(y_{02} - y_{00}y_{12} - y_{01}y_{22})} = \pm r, \; y_{00} - s = x_{01}/(2r^2), \; y_{00} + s = -x_{21}/2,$$

and s and y_{00} are determined uniquely. Further, from

$$y_{12} - x_{01}y_{22}/r^2 = (x_{02} \pm r/2)/r^2,$$

$$y_{12} + x_{21}y_{22} = x_{22} \pm 1/(2r),$$

two pairs of (y_{12}, y_{22}) are obtained (noting that $x_{21} + x_{01}/r^2 \neq 0$). Finally, from

$$y_{02} - y_{00}y_{12} - y_{01}y_{22} = \pm s/2r$$

y_{02} can be obtained. Therefore, every element in (V_2) and two elements in (IV_3) are in the same coset of $PO_3(\mathbb{F}_q)$. □

LEMMA 5.11

For every element in (V_3), there exist another element in (V_3) and $q - 1$ elements in (VI_7), such that these $q + 1$ elements are in the same coset of $PO_3(\mathbb{F}_q)$.

PROOF   It has been proved in Lemma 5.10 that every element in (V_3) with another element in (V_3) are in the same coset of $PO_3(\mathbb{F}_q)$. We will use Equation (5.5), and place the matrix in its right-hand side in (V_3) and the matrix in its left-hand side in (VI_7). Suppose the element $-y_{00}$ in Equation (5.5) is nonsquare in $\mathbb{F}_q^*$. If the matrix on right-hand side of Equation (5.5) belongs to (V_3), then

$$t_1 t_2 = -y_{00}, \qquad t_1 + t_2 = 2(y_{00}y_{21} - y_{01})$$

has a solution (t_1, t_2), $t_1 \neq t_2$. This gives $(y_{00}y_{21} - y_{01})^2 + y_{00} = s^2$, $s \in \mathbb{F}_q^*$. Let $r = y_{00}y_{21} - y_{01}$, then $y_{00} = s^2 - r^2$ and $s(r - s) \neq 0$. Put

$$t_1 = r + s, \qquad t_2 = r - s,$$

$$a = \frac{y_{02} + ry_{12} + (r^2 - s^2)y_{22}}{s(s - r)},$$

$$\lambda = \frac{s(s - r)}{2(y_{02} + ry_{12} + (r^2 - s^2)y_{22})^2}.$$

{Here $y_{02} + ry_{12} + (r^2 - s^2)y_{22} = y_{02} + y_{12}(y_{00}y_{21} - y_{01}) - y_{00}y_{22} \neq 0$.} The right-hand side of Equation (5.5) becomes

$$\begin{pmatrix} \frac{s^2(s^2-r^2)}{(y_{02}+ry_{12}+(r^2-s^2)y_{22})^2} & \frac{s^2(s^2-r^2)\left(y_{21}+(2(r+s))^{-1}\right)}{\left(y_{02}+ry_{12}+(r^2-s^2)y_{22}\right)^2} & \frac{s(s-r)\left(y_{02}+(r+s)y_{12}+(r+s)^2y_{22}\right)}{2\left(y_{02}+ry_{12}+(r^2-s^2)y_{22}\right)^2} \\ 0 & 0 & 1 \\ 1 & y_{21} - \frac{1}{2(s-r)} & \frac{y_{02}+(r-s)y_{12}+(r-s)^2y_{22}}{2s(s-r)} \end{pmatrix}$$
(5.18)

which is in (V_3). For a given element in (V_3),

$$\begin{pmatrix} x_{00} & x_{01} & x_{02} \\ 0 & 0 & 1 \\ 1 & x_{21} & x_{22} \end{pmatrix}$$
(5.19)

where $-x_{00} \in \mathbb{F}_q^*$ is nonsquare and $x_{00}x_{21} - x_{01} \neq 0$. Assume Equation (5.18) equals Equation (5.19). If x_{00} has a representation as

$$x_{00} = u^2 - v^2, \quad u, v \in \mathbb{F}_q$$
(5.20)

Since $-x_{00}$ is nonsquare, so $u \neq 0$. Let

$$\frac{s^2}{y_{02} + ry_{12} + (r^2 - s^2)y_{22}} = u, \qquad \frac{rs}{y_{02} + ry_{12} + (r^2 - s^2)y_{22}} = v$$

then $r/s = v/u$. As

$$y_{21} - \frac{1}{2(s - r)} = x_{21}, \qquad y_{21} + \frac{1}{2(s - r)} = \frac{x_{01}}{x_{00}},$$

it is obtained that

$$s = \left(1 - \frac{v}{u}\right)^{-1}\left(\frac{x_{01}}{x_{00}} - x_{21}\right)^{-1}, \quad r = \frac{vs}{u}, \; y_{21} = x_{21} + \frac{u}{2s(u - v)}.$$

So

$$y_{01} = (s^2 - r^2)\left(x_{21} + \frac{u}{2s(u - v)}\right) - r.$$

Furthermore, y_{02}, y_{12}, y_{22} can be obtained by

$$\begin{pmatrix} y_{02} \\ y_{12} \\ y_{22} \end{pmatrix} = (\lambda \cdot R(a, t_1, t_2))^{-1}\begin{pmatrix} x_{00} & x_{01} & x_{02} \\ 0 & 0 & 1 \\ 1 & x_{21} & x_{22} \end{pmatrix}.$$

It is easy to see that

$$-y_{00} = r^2 - s^2 = s^{-2}\left((rs)^2 - s^4\right) = s^{-2}(v^2 - u^2)$$
$$= -x_{00}s^{-2}\left(y_{02} + ry_{12} + (r^2 - s^2)y_{22}\right)^2,$$

and that it is nonsquare in $\mathbb{F}_q^*$. Therefore, by Equation (5.5), for a given representation of x_{00} in Equation (5.20), if $v \neq 0$, we can find two elements in (VI$_7$) (take r and $-r$ respectively) which are in the same coset of $PO_3(\mathbb{F}_q)$ with Equation (5.19); when $v = 0$ (and $r = 0$ in this case), we can find one element in (VI$_7$) which is in the same coset of $PO_3(\mathbb{F}_q)$ with Equation (5.19).

When $q \equiv 1 \,(mod\, 4)$, $-x_{00}$ and x_{00} are both nonsquare. From Lemma 5.4, there are $(q-1)/4$ pairs $(u^2, v^2)\,(u \neq 0, v \neq 0)$, such that $x_{00} = u^2 - v^2$. Hence, we can find $(q-1)/2$ elements in (VI$_7$) that are in the same coset of $PO_3(\mathbb{F}_q)$ with Equation (5.19). When $q \equiv 3 \,(mod\, 4)$, x_{00} is a square, that is, $x_{00} = u^2$. There are $(q-3)/4$ pairs $(u^2, v^2)\,(u \neq 0, v \neq 0)$, such that $x_{00} = u^2 - v^2$. As above we can find $(q-1)/2$ elements in (VI$_7$) that are in the same coset of $PO_3(\mathbb{F}_q)$ with Equation (5.19).

Every element in (V$_3$) is in the same coset of $PO_3(\mathbb{F}_q)$ with another element in (V$_3$). From Lemma 5.10, their corresponding values of $x_{01} - x_{00}x_{21}$ are different by a minus. Therefore, the elements in (VI$_7$) obtained by the above construction in one coset of $PO_3(\mathbb{F}_q)$ are all different. $\square$

From the above Lemmas 5.5 through 5.11 the following theorem follows.

THEOREM 5.4

Suppose that q is an odd integer. The set of the following $q^2(q^3 - 1)$ elements consists of a representation system of the cosets of $PO_3(\mathbb{F}_q)$ in $PGL_3(\mathbb{F}_q)$.

(i)
$$\begin{pmatrix} x_{00} & x_{01} & x_{02} \\ 0 & 1 & 0 \\ 0 & 0 & 1 \end{pmatrix} \quad x_{00} \neq 0,$$

(ii)
$$\begin{pmatrix} x_{00} & x_{01} & x_{02} \\ 1 & x_{11} & 0 \\ 0 & 0 & 1 \end{pmatrix} \quad x_{00}x_{11} - x_{01} \neq 0,$$

(iii)
$$\begin{pmatrix} x_{00} & x_{01} & x_{02} \\ 0 & 0 & 1 \\ 0 & 1 & x_{22} \end{pmatrix} \quad x_{00} \neq 0,$$

(iv)
$$\begin{pmatrix} x_{00} & -x_{00}^2 + k & x_{02} \\ 1 & 0 & x_{12} \\ 0 & 1 & x_{22} \end{pmatrix} \quad x_{02} - x_{00}x_{12} + (x_{00}^2 - k)x_{22} \neq 0,$$

where $k \in \mathbb{F}_q^*$ is nonsquare, and one of the following two

$$\begin{pmatrix} x_{00} & -x_{00}^2 + k\,x_{02} \\ 1 & 0 & x_{12} \\ 0 & 1 & x_{22} \end{pmatrix}, \quad \begin{pmatrix} x_{00} & -x_{00}^2 + k\,x_{02}' \\ 1 & 0 & x_{12}' \\ 0 & 1 & x_{22}' \end{pmatrix}$$

is taken, where

$$\begin{pmatrix} x_{02}' \\ x_{12}' \\ x_{22}' \end{pmatrix} = \begin{pmatrix} x_{00}^2/k & x_{00}(1 - x_{00}^2/k) & k(1 - x_{00}^2/k)^2 \\ 2x_{00}/k & 1 - 2x_{00}^2/k & -2x_{00}(1 - x_{00}^2/k) \\ 1/k & -x_{00}/k & x_{00}^2/k \end{pmatrix} \begin{pmatrix} x_{02} \\ x_{12} \\ x_{22} \end{pmatrix}.$$

(v)
$$\begin{pmatrix} x_{00} & x_{01} & x_{02} \\ 0 & 0 & 1 \\ 1 & x_{21} & x_{22} \end{pmatrix} \quad x_{00} \neq 0, \quad x_{01} - x_{00}x_{21} \neq 0,$$

only one of the following two elements

$$\begin{pmatrix} x_{00} & x_{01} & x_{02} \\ 0 & 0 & 1 \\ 1 & x_{21} & x_{22} \end{pmatrix}, \quad \begin{pmatrix} x_{00} & x_{00}x_{21} & x_{00}x_{22} \\ 0 & 0 & 1 \\ 1 & x_{01}/x_{00} & x_{02}/x_{00} \end{pmatrix}$$

is taken.

(vi)
$$\begin{pmatrix} x_{00} & x_{01} & x_{02} \\ 0 & 1 & x_{12} \\ 0 & x_{21} & x_{22} \end{pmatrix} \quad x_{00}x_{22} + x_{01}x_{12} - x_{02} - x_{00}x_{12}x_{21} \neq 0,$$

here $-x_{00} \in \mathbb{F}_q^*$ is nonsquare, and there does not exist (r, s) such that $x_{00} = s^2 - r^2, x_{01} = x_{00}x_{21} - r$. Taking $q^3(q-1)^2/4$ elements from this set, any two of them belong to different cosets in $PO_3(\mathbb{F}_q)$. The sizes of the above six sets are $q^2(q-1), q^3(q-1), q^3(q-1), q^3(q-1)^2/4, q^3(q-1)^2/2,$ and $q^3(q-1)^2/4,$ respectively.

Note: It is an open problem how to choose $q^3(q-1)^2/4$ elements that consist of a representation system of the cosets for (vi).

5.4   Encoding Rules $(n = 2, q$ Even$)$

In this section we consider the case of $n = 2$ and q is even. Similar to Section 5.3, we will first show a representation system of the cosets of the subgroup $T = T_3$ in $PGL_3(\mathbb{F}_q)$, and then show a representation system of the cosets of $PO_3(\mathbb{F}_q)$ in $PGL_3(\mathbb{F}_q)$. The group $PO_3(\mathbb{F}_q)$ consists of the following three classes of elements:

$$T(a,t) = \begin{pmatrix} 1 & t & t^2 \\ 0 & a & 0 \\ 0 & 0 & a^2 \end{pmatrix}, Q(a,t) = \begin{pmatrix} 0 & 0 & a^2 \\ 0 & a & 0 \\ 1 & t & t^2 \end{pmatrix}, R(a,t_1,t_2) = \begin{pmatrix} 1 & t_1 & t_1^2 \\ 0 & (t_1+t_2)a & 0 \\ a^2 & t_2a^2 & t_2^2a^2 \end{pmatrix}.$$

Here, $t, t_1, t_2 \in \mathbb{F}_q$, $t_1 \neq t_2$, $a \in \mathbb{F}_q^*$.

Theorem 5.3 cannot be applied when we consider the representation system of cosets of T, as it was assumed in Theorem 5.3 that n and q are relatively prime. The representation system of cosets of T will be shown in the following Theorem 5.5 when q is even. Since q is even, $\psi : t \longmapsto t + t^2$ is a homomorphic of $\mathbb{F}_q$, the kernel of ψ is $\{0, 1\}$. So $|Im(\psi)| = q/2$. Take an element $\alpha \in \mathbb{F}_q$, $\alpha \notin Im(\psi)$. In the following, α denotes such a fixed element.

THEOREM 5.5

Suppose q is an even integer. The following $q^2(q+1)(q^3-1)$ transformations form a representation system of the cosets of T in $PGL_3(\mathbb{F}_q)$.

(I)
$$\begin{pmatrix} x_{00} & 0 & x_{02} \\ 0 & 0 & 1 \\ 0 & 1 & x_{22} \end{pmatrix}, \quad x_{00} \neq 0,$$

(II)
$$\begin{pmatrix} x_{00} & 0 & x_{02} \\ 0 & 1 & x_{12} \\ 0 & 0 & 1 \end{pmatrix}, \quad x_{00} \neq 0,$$

(III)
$$\begin{pmatrix} 0 & x_{01} & x_{02} \\ 1 & x_{11} & x_{12} \\ 0 & 0 & 1 \end{pmatrix}, \quad x_{01} \neq 0,$$

(IV)
$$\begin{pmatrix} x_{00} & 0 & x_{02} \\ 0 & 1 & x_{12} \\ 0 & 1 & x_{22} \end{pmatrix}, \quad x_{00}(x_{12}+x_{22}) \neq 0,$$

$$\begin{pmatrix} x_{00} & \alpha & x_{02} \\ 0 & 1 & x_{12} \\ 0 & 1 & x_{22} \end{pmatrix}, \quad x_{00}(x_{12}+x_{22}) \neq 0$$

where x_{02} runs over a representation system of $\mathbb{F}_q/(x_{12}+x_{22})$ and $\mathbb{F}_q$ is regarded as an additive group.

(V)
$$\begin{pmatrix} 0 & x_{01} & x_{02} \\ 0 & 0 & 1 \\ 1 & x_{21} & x_{22} \end{pmatrix}, \quad x_{01} \neq 0,$$

(VI) $\qquad \begin{pmatrix} 0 & x_{01} & x_{02} \\ 0 & 1 & x_{12} \\ 1 & x_{21} & x_{22} \end{pmatrix}$, $\quad x_{02} + x_{01}x_{12} \neq 0$,

(VII) $\qquad \begin{pmatrix} 0 & x_{01} & x_{02} \\ 1 & x_{11} & x_{12} \\ 0 & 1 & x_{22} \end{pmatrix}$, $\quad x_{02} + x_{01}x_{22} \neq 0$,

(VIII) $\qquad \begin{pmatrix} 0 & x_{01} & x_{02} \\ 1 & x_{11} & x_{12} \\ 1 & x_{21} & x_{22} \end{pmatrix}$, $\quad x_{01}x_{12} + x_{02}x_{21} + x_{02}x_{11} + x_{01}x_{22} \neq 0$,

$$\begin{pmatrix} \alpha & x_{01} & x_{02} \\ 1 & x_{11} & x_{12} \\ 1 & x_{21} & x_{22} \end{pmatrix}, \quad \begin{array}{l} \alpha x_{11}x_{22} + x_{01}x_{12} + x_{02}x_{21} \\ +x_{02}x_{11} + x_{01}x_{22} + \alpha x_{12}x_{21} \neq 0 \end{array}$$

where x_{01} runs over a representation system of $\mathbb{F}_q/(x_{11}+x_{21})$ when $x_{11} \neq x_{21}$; x_{02} runs over a representation system of $\mathbb{F}_q/(x_{12} + x_{22})$ when $x_{11} = x_{21}$. The sizes of the above eight families are $q^2(q-1), q^2(q-1), q^3(q-1), q^2(q-1)^2, q^3(q-1), q^4(q-1), q^4(q-1), q^3(q+1)(q-1)^2$, respectively.

PROOF   Since

$$\frac{|PGL_3(\mathbb{F}_q)|}{|T|} = q^2(q+1)(q^3-1),$$

it is just the sum of sizes of the above eight families. So we only need to prove that any two elements from the above eight families belong to different cosets of T.

Let $Y = (y_{ij}) \in PGL_3(\mathbb{F}_q)$. Then the last two rows of $T(a,t)Y$ are

$$\begin{array}{ccc} ay_{10} & ay_{11} & ay_{12} \\ a^2y_{20} & a^2y_{21} & a^2y_{22} \end{array}$$

Therefore, on the last two rows of Y and $T(a,t)Y$, zero appears in the same positions, and hence, two elements from two families, respectively, belong to different cosets of T.

Suppose Y and $T(a,t)Y$ belong to the same families. It is obtained that $a = 1$ by comparing their third rows. The first row of $T(1,t)Y$ is

$$y_{00} + ty_{10} + t^2y_{20} \quad y_{01} + ty_{11} + t^2y_{21} \quad y_{02} + ty_{12} + t^2y_{22}.$$

It is then obtained that $t = 0$ when Y and $T(1,t)Y$ belong to (I), (II), (III), (V), (VI), or (VII). This shows that two elements from one of these six families belong to different cosets. Suppose Y and $T(1,t)Y$ belong to (IV). Then the first row of $T(1,t)Y$ is

$$y_{00} \quad t+t^2 \quad y_{02}+ty_{12}+t^2y_{22}$$

or

$$y_{00} \quad \alpha+t+t^2 \quad y_{02}+ty_{12}+t^2y_{22}.$$

If Y and $T(1,t)Y$ both have the first form in (IV), then $t+t^2=0$, it deduces that $t=0$ or 1. As y_{02} runs over a fixed representation system of $\mathbb{F}_q/(y_{12}+y_{22})$, therefore, $t=0$ and $Y=T(1,t)Y$. If Y and $T(1,t)Y$ both have the second form of (IV), it is also obtained that $t=0$. For any $t \in \mathbb{F}_q$, $\alpha+t+t^2$ is not zero, so Y and $T(1,t)Y$ cannot have different forms in (IV). This proves that any two elements in (IV) are in different cosets of T. Similarly, the same conclusion is true for the family (VIII). $\qquad\square$

Note that

$$|PO_3(\mathbb{F}_q)/T| = q+1.$$

In order to find a representation system of $PO_3(\mathbb{F}_q)$, we partition the representations of T, obtained in Theorem 5.5, into $q^2(q^3-1)$ parts each consisting of $q+1$ elements, which are in the same coset of $PO_3(\mathbb{F}_q)$.

Partition (VI) into two subsets:

(VI$_1$)
$$\begin{pmatrix} 0 & 0 & x_{02} \\ 0 & 1 & x_{12} \\ 1 & x_{21} & x_{22} \end{pmatrix}, \quad x_{02} \neq 0,$$

(VI$_2$)
$$\begin{pmatrix} 0 & x_{01} & x_{02} \\ 0 & 1 & x_{12} \\ 1 & x_{21} & x_{22} \end{pmatrix}, \quad x_{01} \neq 0, x_{02}+x_{01}x_{12} \neq 0,$$

Then

$$|VI_1| = q^3(q-1), \qquad |VI_2| = q^3(q-1)^2.$$

Partition (VII) into two subsets:

(VII$_1$)
$$\begin{pmatrix} 0 & 0 & x_{02} \\ 1 & x_{11} & x_{12} \\ 0 & 1 & x_{22} \end{pmatrix}, \quad x_{02} \neq 0,$$

(VII$_2$)
$$\begin{pmatrix} 0 & x_{01} & x_{02} \\ 1 & x_{11} & x_{12} \\ 0 & 1 & x_{22} \end{pmatrix}, \quad x_{01} \neq 0, x_{02}+x_{01}x_{22} \neq 0,$$

Then

$$|VII_1| = q^3(q-1), \qquad |VII_2| = q^3(q-1)^2.$$

Partition (VIII) into three subsets:

(VIII$_1$)
$$\begin{pmatrix} 0 & 0 & x_{02} \\ 1 & x_{11} & x_{12} \\ 1 & x_{21} & x_{22} \end{pmatrix}, \quad x_{02}(x_{21} + x_{11}) \neq 0,$$

(VIII$_2$)
$$\begin{pmatrix} 0 & x_{01} & x_{02} \\ 1 & x_{11} & x_{12} \\ 1 & x_{21} & x_{22} \end{pmatrix}, \quad x_{01} \neq 0, \ x_{01}(x_{12} + x_{22}) + x_{02}(x_{11} + x_{21}) \neq 0,$$

(VIII$_3$)
$$\begin{pmatrix} \alpha & x_{01} & x_{02} \\ 1 & x_{11} & x_{12} \\ 1 & x_{21} & x_{22} \end{pmatrix}, \quad \begin{array}{l} \alpha(x_{11}x_{22} + x_{12}x_{21}) + \\ x_{01}(x_{12} + x_{22}) + x_{02}(x_{11} + x_{21}) \neq 0, \end{array}$$

Then

$$|\text{VIII}_1| = q^3(q-1)^2, \ |\text{VIII}_2| = q^3(q-1)^3/2, \ |\text{VIII}_3| = q^3(q+1)(q-1)^2/2.$$

LEMMA 5.12
For every element in (I), *there are* q *elements in* (V) *such that these* $q + 1$ *elements are in the same coset of* $PO_3(\mathbb{F}_q)$.

PROOF The product of $\lambda \cdot R(a, t_1, t_2)$ with an element $Y = (y_{ij})$ in (V) is

$$\lambda \cdot R(a, t_1, t_2) \begin{pmatrix} 0 & y_{01} & y_{02} \\ 0 & 0 & 1 \\ 1 & y_{21} & y_{22} \end{pmatrix}$$
$$= \lambda \begin{pmatrix} t_1^2 & y_{01} + t_1^2 y_{21} & y_{02} + t_1 + t_1^2 y_{22} \\ 0 & 0 & (t_1 + t_2)a \\ t_2^2 a^2 & a^2 y_{01} + a^2 t_2^2 y_{21} & a^2 y_{02} + a^2 t_2 + a^2 t_2^2 y_{22} \end{pmatrix} \tag{5.21}$$

where $y_{01} \neq 0$.

If $y_{21} \neq 0$, let $\lambda = y_{21}$, $t_2 = 0$, $t_1 = \sqrt{y_{01} y_{21}^{-1}}$, $a = \sqrt{y_{01}^{-1} y_{21}^{-1}}$ (as q is even, every element in $\mathbb{F}_q$ has a unique square root). The right-hand side of Equation (5.21) becomes

$$\begin{pmatrix} y_{01} & 0 & y_{21} y_{02} + y_{22} y_{01} + \sqrt{y_{01} y_{21}} \\ 0 & 0 & 1 \\ 0 & 1 & y_{02} y_{01}^{-1} \end{pmatrix}$$

which is in (I). Using

$$y_{01} = x_{00}, \quad y_{21} y_{02} + y_{22} y_{01} + \sqrt{y_{01} y_{21}} = x_{02}, \quad y_{02} y_{01}^{-1} = x_{22}$$

we have

$$y_{01} = x_{00}, \quad y_{02} = x_{00}x_{22}, \quad y_{22} = \left(x_{02} + y_{21}x_{00}x_{22} + \sqrt{y_{21}x_{00}}\right)/x_{00}.$$

Here, y_{21} can be any element of $\mathbb{F}_q^*$.
   If $y_{21} = 0$, we have

$$y_{01}^{-1}Q(y_{01}, 0)\begin{pmatrix} 0 & y_{01} & y_{02} \\ 0 & 0 & 1 \\ 1 & 0 & y_{22} \end{pmatrix} = \begin{pmatrix} y_{01} & 0 & y_{01}y_{22} \\ 0 & 0 & 1 \\ 0 & 1 & y_{02}y_{01}^{-1} \end{pmatrix}$$

The element in the right-hand side is in (I). From $y_{01} = x_{00}$, $y_{01}y_{22} = x_{02}$, $y_{02}y_{01}^{-1} = x_{22}$ it is obtained that $y_{01} = x_{00}$, $y_{02} = x_{00}x_{22}$, $y_{22} = x_{02}x_{00}^{-1}$. Thus, we have proved that every element in (I) and q elements in (V) belong to the same coset of $PO_3(\mathbb{F}_q)$. $\quad\Box$

LEMMA 5.13
For every element in (II), *there are q elements in* (VI$_1$) *such that these $q+1$ elements are in the same coset of $PO_3(\mathbb{F}_q)$.*

PROOF   The product of $\lambda \cdot R(a, t_1, t_2)$ with an element $Y = (y_{ij})$ in (VI$_1$) is

$$\lambda\cdot R(a,t_1,t_2)\begin{pmatrix} 0 & 0 & y_{02} \\ 0 & 1 & y_{12} \\ 1 & y_{21} & y_{22} \end{pmatrix} = \lambda \begin{pmatrix} t_1^2 & t_1 + t_1^2 y_{21} & y_{02} + t_1 y_{12} + t_1^2 y_{22} \\ 0 & (t_1 + t_2)a & (t_1 + t_2)ay_{12} \\ t_2^2 a^2 & (t_2 + t_2^2 y_{21})a^2 & (y_{02} + t_2 y_{12} + t_2^2 y_{22})a^2 \end{pmatrix},$$

where $y_{02} \neq 0$.
   If $y_{21} \neq 0$, let $\lambda = y_{02}y_{21}^2$, $t_2 = 0$, $t_1 = y_{21}^{-1}$, $a = y_{02}^{-1}y_{21}^{-1}$, then the right-hand side becomes

$$\begin{pmatrix} y_{02} & 0 & y_{02}^2 y_{21}^2 + y_{02}y_{12}y_{21} + y_{02}y_{22} \\ 0 & 1 & y_{12} \\ 0 & 0 & 1 \end{pmatrix}.$$

It belongs to (II). Suppose it is the following element in (II)

$$\begin{pmatrix} x_{00} & 0 & x_{02} \\ 0 & 1 & x_{12} \\ 0 & 0 & 1 \end{pmatrix} \quad x_{00} \neq 0,$$

then

$$y_{02} = x_{00}, \; y_{12} = x_{12}, \; y_{22} = \left(x_{02} + x_{00}y_{21}^2 + x_{00}x_{12}y_{21}\right)/x_{00}.$$

Here, y_{21} is an arbitrary element in $\mathbb{F}_q^*$.

If $y_{21} = 0$, we have

$$y_{02}^{-1}Q(y_{02}, 0) \begin{pmatrix} 0 & 0 & y_{02} \\ 0 & 1 & y_{12} \\ 1 & 0 & y_{22} \end{pmatrix} = \begin{pmatrix} y_{02} & 0 & y_{02}y_{22} \\ 0 & 1 & y_{12} \\ 0 & 0 & 1 \end{pmatrix},$$

the element in the right-hand side is in (II). So every element in (II) is in the same coset of $PO_3(\mathbb{F}_q)$ as q elements in (VI$_1$). ▯

LEMMA 5.14

For every element in (III), *there exist one element in* (VII$_1$) *and* $q-1$ *elements in* (VIII$_1$) *such that these* $q + 1$ *elements are in the same coset of* $PO_3(\mathbb{F}_q)$.

PROOF The product of an element $\lambda \cdot Q(a, t)$ and an element $Y = (y_{ij})$ in (VII$_1$) is

$$\lambda \cdot Q(a, t) \begin{pmatrix} 0 & 0 & y_{02} \\ 1 & y_{11} & y_{12} \\ 0 & 1 & y_{22} \end{pmatrix} = \lambda \begin{pmatrix} 0 & a^2 & a^2 y_{22} \\ a & ay_{11} & ay_{12} \\ t & ty_{11} + t^2 & y_{02} + ty_{12} + t^2 y_{22} \end{pmatrix},$$

where $y_{02} \neq 0$. Let $\lambda = y_{02}^{-1}$, $t = 0$, $a = y_{02}$, then the element of the right-hand side is

$$\begin{pmatrix} 0 & y_{02} & y_{02}y_{22} \\ 1 & y_{11} & y_{12} \\ 0 & 0 & 1 \end{pmatrix},$$

which belongs to (III). Suppose it equals $X = (x_{ij})$ in (III), then the following unique solution is obtained:

$$y_{02} = x_{02}, \ y_{11} = x_{11}, \ y_{12} = x_{12}, \ y_{22} = x_{02}x_{01}^{-1}.$$

The product of $\lambda \cdot R(a, t_1, t_2)$ with $Y = (y_{ij})$ in (VIII$_1$) is

$$\lambda \cdot R(a, t_1, t_2) \begin{pmatrix} 0 & 0 & y_{02} \\ 1 & y_{11} & y_{12} \\ 1 & y_{21} & y_{22} \end{pmatrix}$$

$$= \lambda \begin{pmatrix} t_1 + t_1^2 & t_1 y_{11} + t_1^2 y_{21} & y_{02} + t_1 y_{12} + t_1^2 y_{22} \\ (t_1 + t_2)a & (t_1 + t_2)ay_{11} & (t_1 + t_2)ay_{12} \\ (t_2 + t_2^2)a^2 & (t_2 y_{11} + t_2^2 y_{21})a^2 & (y_{02} + t_2 y_{12} + t_2 y_{22}^2)a^2 \end{pmatrix},$$

where $y_{02}(y_{11} + y_{21}) \neq 0$. Let $\lambda = y_{02}$, $t_2 = 0$, $t_1 = 1$, $a = y_{02}^{-1}$, the element in the right-hand side is

$$\begin{pmatrix} 0 & y_{02}(y_{11} + y_{21}) & y_{02}(y_{02} + y_{12} + y_{22}) \\ 1 & y_{11} & y_{12} \\ 0 & 0 & 1 \end{pmatrix}.$$

It belongs to (III). Suppose it is written as $X = (x_{ij})$ in (III), then it is obtained that

$$y_{11} = x_{11}, \quad y_{12} = x_{12}, \quad y_{21} = x_{01}y_{02}^{-1} + x_{11}, \quad y_{22} = x_{02}y_{02}^{-1} + y_{02} + x_{12},$$

where y_{02} ia an arbitrary element in $\mathbb{F}_q^*$. So every element in (III) and one element in (VII$_1$), and $q - 1$ elements in (VIII$_1$) are in the same coset of $PO_3(\mathbb{F}_q)$. $\Box$

LEMMA 5.15

For any element in (IV), there exist q elements in (VI$_2$) such that these $q+1$ elements are in the same coset of $PO_3(\mathbb{F}_q)$.

PROOF   First, we prove that for every element in (VI$_2$), there exist other $q - 1$ elements in (VI$_2$), such that they are in the same coset of $PO_3(\mathbb{F}_q)$. We have

$$\lambda \cdot R(a,t_1,t_2) \begin{pmatrix} 0 & y_{01} & y_{02} \\ 0 & 1 & y_{12} \\ 1 & y_{21} & y_{22} \end{pmatrix}$$
$$= \lambda \begin{pmatrix} t_1^2 & y_{01} + t_1 + t_1^2 y_{21} & y_{02} + t_1 y_{12} + t_1^2 y_{22} \\ 0 & (t_1 + t_2)a & (t_1 + t_2)ay_{12} \\ t_2^2 a^2 & a^2(y_{01} + t_2 + t_2^2 y_{21}) & a^2(y_{02} + t_2 y_{12} + t_2^2 y_{22}) \end{pmatrix}. \tag{5.22}$$

Here, $Y = (y_{ij})$ belongs to (VI$_2$). That is, $y_{01} \neq 0$, $y_{02} + y_{01}y_{12} \neq 0$. Let $\lambda = 1$, $t_2 = a^{-1}$, $t_1 = 0$, then the right-hand side is

$$\begin{pmatrix} 0 & y_{01} & y_{02} \\ 0 & 1 & y_{12} \\ 1 & a^2 y_{01} + a + y_{21} & a^2 y_{02} + ay_{12} + y_{22} \end{pmatrix}.$$

If

$$a^2 y_{01} + a + y_{21} = b^2 y_{01} + b + y_{21},$$
$$a^2 y_{02} + ay_{12} + y_{22} = b^2 y_{02} + by_{12} + y_{22},$$

then either $a = b$ or $a - b = y_{01}^{-1} = y_{12}y_{02}^{-1}$ ($y_{02} \neq 0$ in this case), but $y_{02} + y_{01}y_{12} \neq 0$. Therefore $a = b$, a is an arbitrary element in $\mathbb{F}_q^*$. This shows that every element in (VI$_2$) with other $q - 1$ elements in (VI$_2$) are in the same coset of $PO_3(\mathbb{F}_q)$.

Let $\lambda = y_{01}$, $t_1 = 1$, $t_2 = 0$, $a = y_{01}^{-1}$, then the right-hand side of Equation (5.22) becomes

$$\begin{pmatrix} y_{01} & y_{01}^2 + y_{01} + y_{21} & y_{01}y_{02} + y_{01}y_{12} + y_{01}y_{22} \\ 0 & 1 & y_{12} \\ 0 & 1 & y_{02}y_{01}^{-1} \end{pmatrix},$$

which belongs to (IV). Suppose it equals the following element of (IV):

$$\begin{pmatrix} x_{00} & u & x_{02} \\ 0 & 1 & x_{12} \\ 0 & 1 & x_{22} \end{pmatrix} \qquad u \in \{0, \alpha\}.$$

Then we have the solution

$$y_{01} = x_{00}, \qquad y_{02} = x_{00}x_{22}, \qquad y_{12} = x_{12},$$

$$y_{21} = x_{00} + 1 + ux_{00}^{-1},$$

$$y_{22} = x_{02}x_{00}^{-1} + x_{00}x_{22} + x_{12}.$$

This proves that every element in (IV) with q elements of (VI$_2$) is in the same coset of $PO_3(\mathbb{F}_q)$. □

LEMMA 5.16
For every element in (VII$_2$), *there exist another element in* (VII$_2$) *and* $q - 1$ *elements in* (VIII$_2$) *such that they are in the same coset of* $PO_3(\mathbb{F}_q)$.

PROOF   It is known that

$$x_{01}^{-1} \cdot Q(x_{01}, 0) \begin{pmatrix} 0 & x_{01} & x_{02} \\ 1 & x_{11} & x_{12} \\ 0 & 1 & x_{22} \end{pmatrix} = \begin{pmatrix} 0 & x_{01} & x_{01}x_{22} \\ 1 & x_{11} & x_{12} \\ 0 & 1 & x_{02}x_{01}^{-1} \end{pmatrix}.$$

If $x_{01} \neq 0$, $x_{02} \neq x_{01}x_{22}$, from the above we know that every element in (VII$_2$) with another element in (VII$_2$) are in the same coset of $PO_3(\mathbb{F}_q)$.

As

$$x_{01} \cdot R(x_{01}^{-1}, 0, 0) \cdot \begin{pmatrix} 0 & x_{01} & x_{01}x_{22} \\ 1 & x_{11} & x_{12} \\ 1 & 1 + x_{01} + x_{11} & x_{02}x_{01}^{-1} + x_{22}x_{01} + x_{12} \end{pmatrix}$$

$$= \begin{pmatrix} 0 & x_{01} & x_{02} \\ 1 & x_{11} & x_{12} \\ 0 & 1 & x_{12} \end{pmatrix},$$

this shows that every element in (VII$_2$) with one element in (VIII$_2$) are in the same coset of $PO_3(\mathbb{F}_q)$.

Next we show that every element in (VIII$_2$) with other $q - 2$ elements in (VIII$_2$) are in the same coset of $PO_3(\mathbb{F}_q)$. We have

$$\lambda \cdot R(a, t_1, t_2) \begin{pmatrix} 0 & y_{01} & y_{02} \\ 1 & y_{11} & y_{12} \\ 1 & y_{21} & y_{22} \end{pmatrix} =$$

$$\lambda \cdot \begin{pmatrix} t_1 + t_1^2 & y_{01} + t_1y_{11} + t_1^2y_{21} & y_{02} + t_1y_{12} + t_1^2y_{22} \\ (t_1 + t_2)a & (t_1 + t_2)ay_{11} & (t_1 + t_2)ay_{12} \\ (t_2 + t_2^2)a^2 & (y_{01} + t_2y_{11} + t_2^2y_{21})a^2 & (y_{02} + t_2y_{12} + t^2y_{22})a^2 \end{pmatrix}.$$

If $a \neq 1$, let $\lambda = (1+a)^{-1}$, $t_2 = 1 + a^{-1}$, $t_1 = 0$, then the right-hand side becomes

$$\begin{pmatrix} 0 & (1+a)^{-1}y_{01} & (1+a)^{-1}y_{02} \\ 1 & y_{11} & y_{12} \\ 1 & (1+a)^{-1}a^2 y_{01} + ay_{11} + (1+a)y_{21} & (1+a)^{-1}a^2 y_{02} + ay_{12} + (1+a)y_{22} \end{pmatrix},$$

which belongs to (VIII$_2$). Here, a can be any element of $\mathbb{F}_q \setminus \{0, 1\}$, so Lemma 5.16 is proved. ☐

The following Theorem 5.6 can be proved based on Lemmas 5.12 through 5.16.

THEOREM 5.6

Suppose that q is an even integer. The set of the following $q^2(q^3 - 1)$ elements form a representation system of the cosets of $PO_3(\mathbb{F}_q)$ in $PGL_3(\mathbb{F}_q)$:

(i)
$$\begin{pmatrix} x_{00} & 0 & x_{02} \\ 0 & 0 & 1 \\ 0 & 1 & x_{22} \end{pmatrix}, \quad x_{00} \neq 0,$$

(ii)
$$\begin{pmatrix} x_{00} & 0 & x_{02} \\ 0 & 1 & x_{12} \\ 0 & 0 & 1 \end{pmatrix}, \quad x_{00} \neq 0,$$

(iii)
$$\begin{pmatrix} 0 & x_{01} & x_{02} \\ 1 & x_{11} & x_{12} \\ 0 & 0 & 1 \end{pmatrix}, \quad x_{01} \neq 0,$$

(iv)
$$\begin{pmatrix} 0 & x_{01} & x_{02} \\ 1 & x_{11} & x_{12} \\ 0 & 1 & x_{22} \end{pmatrix}, \quad x_{01}(x_{02} + x_{01}x_{22}) \neq 0,$$

and only one matrix is taken among the following two matrices:

$$\begin{pmatrix} 0 & x_{01} & x_{02} \\ 1 & x_{01} & x_{12} \\ 0 & 1 & x_{22} \end{pmatrix}, \quad \begin{pmatrix} 0 & x_{01} & x_{01}x_{22} \\ 1 & x_{11} & x_{12} \\ 0 & 0 & x_{02}x_{01}^{-1} \end{pmatrix},$$

(v)
$$\begin{pmatrix} x_{00} & 0 & x_{02} \\ 0 & 1 & x_{12} \\ 0 & 1 & x_{22} \end{pmatrix}, \quad x_{00}(x_{12} + x_{22}) \neq 0,$$

$$\begin{pmatrix} x_{00} & \alpha & x_{02} \\ 0 & 1 & x_{12} \\ 0 & 1 & x_{22} \end{pmatrix}, \quad x_{00}(x_{12} + x_{22}) \neq 0,$$

where x_{02} runs over a representation system of $\mathbb{F}_q/(x_{12} + x_{22})$.

(vi)
$$\begin{pmatrix} \alpha & x_{01} & x_{02} \\ 1 & x_{11} & x_{12} \\ 1 & x_{21} & x_{22} \end{pmatrix},$$

where

$$\alpha x_{11}x_{22} + x_{01}x_{12} + x_{02}x_{21} + x_{02}x_{11} + x_{01}x_{22} + \alpha x_{12}x_{21} \neq 0.$$

x_{01} runs over a representation system of $\mathbb{F}_q/(x_{11} + x_{21})$ if $x_{11} \neq x_{21}$, and x_{02} runs over a representation system of $\mathbb{F}_q/(x_{12} + x_{22})$ if $x_{11} = x_{21}$. In this class only $q^3(q-1)^2/2$ elements are taken such that any two of them belong to different cosets of $PO_3(\mathbb{F}_q)$. The sizes of the above six sets are $q^2(q-1), q^2(q-1), q^3(q-1), q^3(q-1)^2/2, q^2(q-1)^2$, and $q^3(q-1)^2/2$, respectively.

Note: It is an open problem how to choose $q^3(q-1)^2/2$ elements that consist of a representation system of the cosets for (vi).

5.5 Comments

Authentication codes based on rational normal curves over finite fields constructed in this chapter are of only one type, known to the author so far as the t-fold perfect non-Cartesian A-code with $t > 2$. The presentation of Lemmas 5.1, 5.2, and 5.3 is standard. Theorem 5.1 is due to the author [30].

All the results of Sections 5.2, 5.3, except Lemma 5.4, are due to the author. Lemma 5.4 is well known. The results of Section 5.4 are due to Pei and Wang [29].

5.6 Exercises

5.1 Calculate the sizes of the families (I) to (VI), and the sizes of the subfamilies (II$_i$) $1 \leq i \leq 2$, (IV$_i$) $1 \leq i \leq 4$, (V$_i$) $1 \leq i \leq 3$, and (VI$_i$) $1 \leq i \leq 8$ of Section 5.3.

5.2 Calculate the sizes of the families (I) to (VIII), and the sizes of the subfamilies (VI$_i$) $1 \leq i \leq 2$, (VII$_i$) $1 \leq i \leq 8$, and (VIII$_i$) $1 \leq i \leq 3$ of Section 5.4.

5.3 Prove that any two elements in the family (VIII) of Section 5.3 belong to different cosets of T.

5.4 Prove that the system of Equations (5.14) has $(q-1)/2$ solutions in $\mathbb{F}_q$ when $k \in \mathbb{F}_q^*$ is nonsquare.

Chapter 6

t-Designs

Let t, v, k, and λ be positive integers. A $t - (v, k, \lambda)$ design is a pair $(\mathcal{M}, \mathcal{E})$ where $\mathcal{M}$ is a set of v elements called points and $\mathcal{E} = (\mathcal{E}_1, \mathcal{E}_2, ..., \mathcal{E}_b)$ is a set of k-subsets of $\mathcal{M}$ — called blocks — such that every t-subset of $\mathcal{M}$ is contained in exactly λ blocks (see Definition 3.5). It was proved in Theorem 3.3 that $t - (v, k, 1)$ designs can be used to construct perfect A-codes with $P_r (0 \leq r < t)$ achieving combinatorial bounds.

We give a basic property of t-designs first.

PROPOSITION 6.1

Let $(\mathcal{M}, \mathcal{E})$ be a $t - (v, k, \lambda)$ design, and $1 \leq r \leq t$. Then $(\mathcal{M}, \mathcal{E})$ is also an $r - (v, k, \lambda_r)$ design, where

$$\lambda_r = \frac{\lambda \begin{pmatrix} v - r \\ t - r \end{pmatrix}}{\begin{pmatrix} k - r \\ t - r \end{pmatrix}}$$

where $\begin{pmatrix} a \\ b \end{pmatrix}$ is the binomial coefficient.

PROOF Suppose that $X \subset \mathcal{M}$ with $|X| = r$. Let $\lambda_r(X)$ denote the number of blocks containing all the points in X. Define the set

$$I = \{(Y, B) \mid Y \subset \mathcal{M}, B \in \mathcal{E}, |Y| = t - r, Y \cap X = \emptyset, X \cup Y \subset B\}.$$

We compute $|I|$ in two different ways.

First, there are $\begin{pmatrix} v - r \\ t - r \end{pmatrix}$ ways to choose Y. For each such Y, there are λ blocks B such that $X \bigcup Y \subset B$. Hence,

$$|I| = \lambda \begin{pmatrix} v - r \\ t - r \end{pmatrix}.$$

On the other hand, there are $\lambda_r(X)$ ways to choose a block B such that

$X \subset B$. For each choice of B, there are $\begin{pmatrix} k-r \\ t-r \end{pmatrix}$ ways to choose Y. Hence,

$$|I| = \lambda_r(X) \begin{pmatrix} k-r \\ t-r \end{pmatrix}.$$

Combining these two equations we see that $\lambda_r(X) = \lambda_r$ as desired. $\quad\square$

The proposition shows that a t-design is always strong. This is a difference between the t-designs and the partially balanced t-designs.

We will present some well-known constructions of $2-(v,k,1)$ and $3-(v,k,1)$ designs in this chapter, and will not consider the t-designs with $t \geq 4$. It is known that such designs exist for arbitrary t [50], but relatively little is known for $t \geq 4$, and their construction is usually complicated.

6.1 $2 - (v, k, 1)$ **Designs**

Let $\mathbb{F}_q$ be the finite field with q elements where q is a prime power, and $AG(2, \mathbb{F}_q)$ be the *affine* plane over $\mathbb{F}_q$. Each point of $AG(2, \mathbb{F}_q)$ is represented by a pair of elements (α, β) from $\mathbb{F}_q$; there are totally q^2 points in $AG(2, \mathbb{F}_q)$. All points (α, β) satisfying the linear equation

$$ax + by + c = 0 \tag{6.1}$$

form a line in $AG(2, \mathbb{F}_q)$ where a, b are not zero simultaneously. It is easy to see that each line has q points. Any two different points of $AG(2, \mathbb{F}_q)$ determines a unique line passing through them, hence the total number of lines in $AG(2, \mathbb{F}_q)$ is

$$\frac{q^2(q^2-1)}{q(q-1)} = q^2 + q,$$

and the total number of lines passing through one fixed point is

$$\frac{q^2-1}{q-1} = q+1.$$

Let $\mathcal{M}$ be the set of all points and $\mathcal{E}$ be the set of all lines in $AG(2, \mathbb{F}_q)$. The above argument shows that the pair $(\mathcal{M}, \mathcal{E})$ is a $2 - (q^2, q, 1)$ design, and it is also a $1 - (q^2, q, q+1)$ design which can also be deduced from Proposition 6.1.

We have introduced the projective plane $PG(2, \mathbb{F}_q)$ in Section 3.4. Each point of $PG(2, \mathbb{F}_q)$ is denoted by (α, β, γ) $(\alpha, \beta, \gamma \in \mathbb{F}_q$, they are not all zero), and $\lambda(\alpha, \beta, \gamma)$ denotes the same point of (α, β, γ) when $\lambda \neq 0$. Hence, the number of points in $PG(2, \mathbb{F}_q)$ is

$$\frac{q^3 - 1}{q - 1} = q^2 + q + 1.$$

All points of $PG(2, \mathbb{F}_q)$ satisfying the linear equation

$$ax + by + cz = 0, \quad a, b, c \in \mathbb{F}_q \tag{6.2}$$

(a, b, c are not all zero) form a line in $PG(2, \mathbb{F}_q)$. Assuming that $a \neq 0$, the coordinate x will be uniquely determined after taking (y, z) as any pair of elements in $\mathbb{F}_q$ except $y = z = 0$. Hence, there are $(q^2 - 1)/(q - 1) = q + 1$ points on every line. Given any two different points in $PG(2, \mathbb{F}_q)$, there is a unique line passing through these two points.

Let $\mathscr{M}$ be the set of all points and $\mathscr{E}$ be the set of all lines in $PG(2, \mathbb{F}_q)$. It is proved by the above calculation that the pair $(\mathscr{M}, \mathscr{E})$ is a $2 - (q^2 + q + 1, q + 1, 1)$ design. Similarly, we can construct $2 - (q^n, q, 1)$ and $2 - ([q^{n+1} - 1]/[q - 1], q + 1, 1)$ designs by using the spaces $AG(n, \mathbb{F}_q)$ and $PG(n, \mathbb{F}_q)$, respectively.

6.2 Steiner Triple System

Let v, k, and λ be positive integers, and $v > k \geq 2$. If the pair $(\mathscr{M}, \mathscr{E})$ has the following properties:

(i) $\mathscr{M}$ is a set of v elements called points,
(ii) $\mathscr{E}$ is a set of nonempty subsets of $\mathscr{M}$ called blocks,
(iii) each block contains exactly k points,
(iv) every pair of distinct points is contained in exactly λ blocks,

then $(\mathscr{M}, \mathscr{E})$ is called a balanced incomplete block design (**BIBD**), denoted by (v, k, λ). The property (iv) is the "balance" property, and the inequality $k < v$ is the "incomplete" property. We assume that there are no repeated blocks in $\mathscr{E}$. (It is called that the design is simple.) For the application to the construction of perfect authentication codes we consider only the case of $\lambda = 1$, i.e., the simple design. In the case of $k = 2$, $\mathscr{E}$ is the set of all blocks with two points, so we do not consider this trivial case. Now we assume that $k = 3$. A $(v, 3, 1)$-BIBD is also called a Steiner Triple System of order v denoted by **STS**(v), which is also a $2 - (v, 3, 1)$ design discussed in the above section.

We discuss the construction of Steiner Triple Systems in this section. We have already seen some Steiner Triple Systems in the above section; the $2 - (7, 3, 1)$ design constructed by using $PG(2, \mathbb{F}_2)$ is a STS(7) and the $2 - (9, 3, 1)$

design constructed by using $AG(2, \mathbb{F}_3)$ is a STS(9). The following are STS(7) and STS(9):

$$\{v = 7\}$$

$$(1, 2, 4),$$
$$(2, 3, 5), \quad (3, 4, 6), \quad (4, 5, 7),$$
$$(5, 6, 1), \quad (6, 7, 2), \quad (7, 1, 3).$$

$$\{v = 9\}$$

$$(1, 2, 3), \quad (4, 5, 6), \quad (7, 8, 9),$$
$$(1, 4, 8), \quad (2, 5, 9), \quad (3, 6, 7),$$
$$(1, 5, 7), \quad (2, 6, 8), \quad (3, 4, 9),$$
$$(1, 6, 9), \quad (2, 4, 7), \quad (3, 5, 8).$$

LEMMA 6.1
In a (v, k, λ)-BIBD, every point occurs in exactly

$$r = \frac{\lambda(v - 1)}{k - 1}$$

blocks.

PROOF   A (v, k, λ)-BIBD is also a $2 - (v, k, \lambda)$ design. By taking $k = 3, t = 2, r = 1$ in Proposition 6.1, the lemma is followed.   ∎

LEMMA 6.2
A (v, k, λ)-BIBD has exactly

$$b = \frac{vr}{k} = \frac{\lambda(v^2 - v)}{k^2 - k}$$

blocks.

PROOF   Let $(\mathscr{M}, \mathscr{E})$ be a (v, k, λ)-BIBD, and let $b = |\mathscr{E}|$. Define a set

$$I = \{(x, B) | x \in \mathscr{M}, B \in \mathscr{E}, x \in B\}.$$

We will compute $|I|$ in two different ways.

First, there are v ways to choose $x \in \mathscr{M}$. For each such x, there are r blocks B such that $x \in B$. Hence,

$$|I| = vr.$$

On the other hand, there are b ways to choose a block $B \in \mathscr{E}$. For each choice of B, there are k ways to choose $x \in B$. Hence,

$$|I| = bk.$$

Combining these two equations, we see that

$$bk = vr$$

as desired. ▯

COROLLARY 6.1
There exists a STS(v) only if $v \equiv 1, 3 \pmod 6$, $v \geq 7$.

PROOF   Suppose there exists a $(v, 3, 1)$-BIBD, we see that

$$r = \lambda(v - 1)/(k - 1) = (v - 1)/2$$

by Lemma 6.1, and

$$b = vr/k = v(v - 1)/6$$

by Lemma 6.2. Since r and b are integers, it follows that v is odd and $v(v-1) \equiv 0 \pmod 6$. Hence, $v \equiv 1, 3 \pmod 6$, and $v \geq 7$ (because $v > k = 3$). ▯

In fact, the conditions in Corollary 6.1 are also sufficient for the existence of STS(v). This conclusion will be proved later for the case $v \equiv 1 \pmod 6$ and the case $v \equiv 3 \pmod 6$, separately.

Let us introduce some concepts first. Suppose "$\circ$" is a binary operation on the set X, i.e., $\circ : X \times X \to X$. If the following conditions are satisfied:
(i) the equation $x \circ z = y$ has a unique solution $z \in X$ for any given $x, y \in X$;
(ii) the equation $z \circ x = y$ has a unique solution $z \in X$ for any given $x, y \in X$ too, then $(X, \circ)$ is called a *pseudo-group*. If $x \circ y = y \circ x$ for any $x, y \in X$, then $(X, \circ)$ is called symmetric, and if $x \circ x = x$ for any $x \in X$, then $(X, \circ)$ is called idempotent. The number $|X|$ is called the order of $(X, \circ)$.

LEMMA 6.3
If there exists a symmetric idempotent pseudo-group of order n, then n is odd.

PROOF   Taking a fixed element $z \in X$, define a set $S = \{(x, y)|x, y \in X, x \circ y = z\}$. For any $x \in X$, there is a unique $y \in X$ such that $x \circ y = z$ by the definition of pseudo-groups, hence $|S| = n$. Let (x, y) be an element of S. If $x = y$ then $x = x \circ x = z$ since $(X, \circ)$ is idempotent, while if $x \neq y$, then $(y, x) \in S$ since $(X, \circ)$ is symmetric. Hence $n - 1$ is even as desired. ▯

Let $n > 0$ be an odd integer. $(\mathbb{Z}_n, +)$ denotes the group with respect to the addition modulo n; it is a symmetric pseudo-group also but it is not idempotent. By taking a permutation π on $\mathbb{Z}_n$, we may obtain an idempotent pseudo-group. The series $\{x + x \pmod n\}$ is $0, 2, 4, ..., n - 1, 1, 3, ..., n - 2$ in

$\mathbb{Z}_n$, while the series $x \circ x$ would be $0, 1, 2, ..., n-1$ when $(\mathbb{Z}_n, \circ)$ becomes an idempotent pseudo-group.

Hence, defining the permutation

$$\pi: \ x \longmapsto \frac{n+1}{2} \cdot x \quad mod \ n$$

and the operation

$$x \circ y = \pi(x+y) = \frac{n+1}{2} \cdot (x+y) \quad mod \ n,$$

then

$$x \circ x = \frac{n+1}{2}(x+x) \equiv x \ mod \ n;$$

$(\mathbb{Z}_n, \circ)$ becomes an idempotent pseudo-group of order n. Note that $2^{-1} \ mod \ n = (n+1)/2$ whenever n is odd.

Bose's Construction

Suppose $v = 6t + 3, t \geq 1$. Let $(\mathbb{Z}_{2t+1}, \circ)$ be the symmetric idempotent pseudo-group of order $2t+1$ constructed above. Define the set $Y = \mathbb{Z}_{2t+1} \times \mathbb{Z}_3$ which will become a point set of a $STS(v)$. For any element $x \in \mathbb{Z}_{2t+1}$ define a block

$$A_x = \{(x, 0), (x, 1), (x, 2)\},$$

and for any two elements $x, y \in \mathbb{Z}_{2t+1}, x < y$ and any $i \in \mathbb{Z}_3$ define a block

$$B_{x,y,i} = \{(x, i), (y, i), (x \circ y, i+1 \ (mod \ 3))\}.$$

Put

$$B = \{A_x, x \in \mathbb{Z}_{2t+1}\} \cup \{B_{x,y,i} : x < y, x, y \in \mathbb{Z}_{2t+1}, i \in \mathbb{Z}_3\}.$$

We will prove (Y, B) is a $SYS(v)$. It is only necessary to prove that any pair of distinct points of Y is contained in a unique block.

Suppose that $(\alpha, j), (\beta, k)$ is an arbitrary pair of elements of Y. If $\alpha = \beta$, then $j \neq k$ and the pair belongs to A_α. If $\alpha \neq \beta$, without loss of generality we may assume that $\alpha < \beta$.

(i) If $j = k$, then (α, j) and (β, k) belong to the unique block $B_{\alpha, \beta, j}$.

(ii) If $k = j + 1 (mod \ 3)$, let γ be the unique element such that $\gamma \circ \alpha = \beta$. We have $\gamma \neq \alpha$, otherwise $\beta = \alpha \circ \alpha = \alpha$. If $\gamma < \alpha$, then (α, j) and (β, k) belong to $B_{\gamma, \alpha, j}$, while if $\gamma > \alpha$, (α, j) and (β, k) belong to $B_{\alpha, \gamma, j}$.

(iii) If $j = k + 1 (mod \ 3)$, the proof is similar to the case of (ii).

This completes the proof.

The displayed STS(9) above is the Bose's construction with $t = 1$. Denote each point of Y by numbers $1, 2, \cdots, 9$:

$$\begin{array}{lll} 1 \leftarrow (0,0) & 2 \leftarrow (0,1) & 3 \leftarrow (0,2) \\ 4 \leftarrow (1,0) & 5 \leftarrow (1,1) & 6 \leftarrow (1,2) \\ 7 \leftarrow (2,0) & 8 \leftarrow (2,1) & 9 \leftarrow (2,2) \end{array}$$

The STS(9) has been constructed.

We have constructed symmetric idempotent pseudo-groups for each odd number. We will construct symmetric semi-idempotent pseudo-groups for each even number in the following. Letting n be even, a pseudo-group $(\mathbb{Z}_n, \circ)$ is called semi-idempotent if

$$x \circ x = \begin{cases} x & if \ 0 \leq x < \frac{n}{2}, \\ x - \frac{n}{2} & if \ \frac{n}{2} \leq x < n. \end{cases}$$

The series $\{x + x \ (mod \ n)\}$ in $\mathbb{Z}_n$ is

$$0, 2, ..., n - 2, 0, 2, ..., n - 2$$

while the series $\{x \circ x\}$ would be

$$0, 1, ..., \frac{n}{2} - 1, 0, 1, ..., \frac{n}{2} - 1$$

when $(\mathbb{Z}_n, \circ)$ is semi-idempotent. Defining the permutation

$$\pi(x) = \begin{cases} \frac{x}{2} & if \ x \ is \ oven \\ \frac{x+n-1}{2} & if \ x \ is \ odd \end{cases}$$

on $\mathbb{Z}_n$, and putting

$$x \circ y = \pi(x + y \ (mod \ n)),$$

it is easy to see that $(\mathbb{Z}_n, \circ)$ becomes a symmetric semi-idempotent pseudo-group.

Skolem's Construction

Let $v = 6t + 1$, $t \geq 1$, and $(\mathbb{Z}_{2t}, \circ)$ be a symmetric semi-idempotent pseudo-group. Put $Y = (\mathbb{Z}_{2t} \times \mathbb{Z}_3) \cup \{\infty\}$. Define a block

$$A_x = \{(x, 0), (x, 1), (x, 2)\}$$

for each $0 \leq x \leq t - 1$, a block

$$B_{x,y,i} = \{(x, i), (y, i), (x \circ y, i + 1)\}$$

for any pair $x, y \in \mathbb{Z}_{2t}$, $x < y$ and $i \in \mathbb{Z}_3$, finally a block

$$C_{x,i} = \{\infty, (x + t, i), (x, i + 1)\}$$

for each $0 \leq x \leq t - 1$ and $i \in \mathbb{Z}_3$. Put

$$B = \{A_x : 0 \leq x \leq t - 1\}$$

$$\cup \{B_{x,y,i} : x, y \in \mathbb{Z}_{2t}, x < y, i \in \mathbb{Z}_3\}$$

$$\cup \{C_{x,i} : 0 \leq x \leq t - 1, i \in \mathbb{Z}_3\}.$$

To prove that (Y, B) is a $STS(v)$, it is necessary only to prove that any pair of distinct elements in Y is contained in exactly one block.

Let (α, j) and ∞ be two elements of Y. If $0 \le \alpha \le t - 1$, they are contained in the unique block $C_{\alpha,j-1}$, while if $\alpha \ge t$ they are contained in the unique block $C_{\alpha-t,j}$.

Let $(\alpha, j), (\beta, k)$ be two elements of Y. If $\alpha = \beta \le t - 1$, then they are contained in the unique block A_α. If $\alpha = \beta \ge t$, then $j \ne k$. Assume that $k \equiv j + 1 \pmod 3$ without loss of generality. Suppose that $x = \gamma$ is the unique solution of $\alpha \circ x = \beta$, then $\alpha \ne \gamma$, otherwise $\beta = \alpha \circ \alpha = \alpha - t < t$. If $\gamma > \alpha$ then $(\alpha, j), (\beta, k)$ are contained in the unique block $B_{\alpha,\gamma,j}$, and if $\gamma < \alpha$ then $(\alpha, j), (\beta, k)$ are contained in the unique block $B_{\gamma,\alpha,j}$, since the operation $\circ$ is symmetric.

Now we consider the case of $\alpha \ne \beta$, and we assume $\alpha < \beta$.

(i) If $j = k$, then $(\alpha, j), (\beta, k)$ are contained in the unique block $B_{\alpha,\beta,j}$.

(ii) If $k \equiv j + 1 \pmod 3$, let γ be the unique solution of $x \circ \alpha = \beta$. Since $\alpha < \beta$, $\alpha \circ \alpha \le \alpha$, we have $\gamma \ne \alpha$. If $\gamma < \alpha$ then $(\alpha, j), (\beta, k)$ are contained in the unique block $B_{\gamma,\alpha,j}$, and if $\gamma > \alpha$ then $(\alpha, j), (\beta, k)$ are contained in the unique block $B_{\alpha,\gamma,j}$.

(iii) If $j \equiv k + 1 \pmod 3$, the proof is similar to that of (ii). This completes the proof.

6.3 $3 - (v, k, 1)$ **Designs**

(I) Let $AG(n, \mathbb{F}_2)$ denote the affine space of dimension n over $\mathbb{F}_2$ and T be an invertible $n \times n$ matrix over $\mathbb{F}_2$. The one-to-one transformation on $AG(n, \mathbb{F}_2)$:

$$AG(n, \mathbb{F}_2) \to AG(n, \mathbb{F}_2)$$

$$(x_1, ..., x_n) \longmapsto (x_1, ..., x_n)T + (c_1, ..., c_n)$$

is called an affine transformation. All affine transformations on $AG(n, \mathbb{F}_2)$ form a group.

Suppose that two vectors $(a_1, ..., a_n)$ and $(b_1, ..., b_n)$ are linearly independent. The set of points

$$s(a_1, ..., a_n) + t(b_1, ..., b_n) + (c_1, ..., c_n), \quad s, t \in \mathbb{F}_2$$

is called a plane of $AG(n, \mathbb{F}_2)$. There are four points on each plane. Planes are mapped to planes under any affine transformation.

LEMMA 6.4

Any three points of $AG(n, \mathbb{F}_2)$ are contained in a unique plane.

PROOF   By taking an affine transformation, if it is necessary, we may assume that these three points are $A = (0, ..., 0)$, $B = (0, ..., 1)$, and $C = (c_1, ..., c_n)$ where $c_1, ..., c_{n-1}$ are not all zero. The fourth point on the plane passing through A, B, and C is $B + C = (c_1, ..., c_{n-1}, c_n + 1)$. Hence, such plane is unique. $\square$

Let $\mathcal{M}$ be the set of all points and $\mathcal{E}$ be the set of all planes in $AG(n, \mathbb{F}_2)$. We have $|\mathcal{M}| = 2^n$. The pair $(\mathcal{M}, \mathcal{E})$ is a $3 - (2^n, 4, 1)$ design as shown by Lemma 6.4. Hence,

$$|\mathcal{E}| = \binom{2^n}{3} / \binom{4}{3} = 2^{n-2}(2^n - 1)(2^{n-1} - 1)/3.$$

The design is also called a Steiner Quadruple System.

Let $\mathcal{M}$ be the set of all $q^n + 1$ points on the projective line $PG(1, \mathbb{F}_{q^n})$, which can be denoted by

$$(1, \alpha), \alpha \in \mathbb{F}_{q^n}; \quad (0, 1).$$

The set $\mathcal{M}$ can be also denoted by $\mathbb{F}_{q^n} \cup \{\infty\}$. Put $B = \mathbb{F}_q \cup \{\infty\}$ which is a block in $\mathcal{M}$ and has $q + 1$ points. Let $G = PGL_2(\mathbb{F}_{q^n})$ be the group of projective transformations of $PG(1, \mathbb{F}_{q^n})$ and let G_B be the subgroup of G which maps the block B into itself, which is isomorphic to the group $PGL_2(\mathbb{F}_q)$ of projective transformations on $PG(1, \mathbb{F}_q)$. Let B^G be the set of all blocks into which the block B is mapped by transformations in G, then

$$|B^G| = |G|/|G_B| = \frac{(q^{2n} - 1)(q^{2n} - q^n)}{q^n - 1} / \frac{(q^2 - 1)(q^2 - q)}{q - 1} = \frac{q^n(q^{2n} - 1)}{q(q^2 - 1)}.$$

Each block in B^G has $q + 1$ points.

LEMMA 6.5
The group $G = PGL_2(\mathbb{F}_{q^n})$ is 3-transitive, i.e., any three points of $PG(1, \mathbb{F}_{q^n})$ can be mapped into any other three points by a transformation in G.

PROOF   It is only necessary to show that any three points $p_i = (a_i, b_i)$, $i = 1, 2, 3$ can be mapped into $(1, 0)$, $(0, 1)$, and $(1, 1)$ by a transformation of G. Any two points from p_i $(1 \leq i \leq 3)$ are linearly independent. Let

$$T_1 = \begin{pmatrix} a_1 & b_1 \\ a_2 & b_2 \end{pmatrix}^{-1},$$

then $(a_1, b_1)T_1 = (1, 0)$, $(a_2, b_2)T_1 = (0, 1)$. Suppose $(a_3, b_3)T_1 = (c, d)$, where c and d are nonzero since p_1, p_3 and p_2, p_3 are linearly independent. Taking

$$T_2 = \begin{pmatrix} c^{-1} & 0 \\ 0 & d^{-1} \end{pmatrix},$$

the projective transformation $T_1 T_2$ maps $p_i (1 \leq i \leq 3)$ into (1,0), (0,1), and (1,1), respectively, as desired.   ▯

Lemma 6.5 shows that each 3-subset of $PG(1, \mathbb{F}_{q^n})$ is at least contained in one block of B^G. The total number of 3-subsets of $PG(1, \mathbb{F}_{q^n})$ is

$$\binom{q^n + 1}{3} = \frac{q^n (q^{2n} - 1)}{6},$$

and the total number of 3-subsets, which is contained in some blocks of B^G, does not exceed

$$|B^G| \cdot \binom{q + 1}{3} = \frac{q^n (q^{2n} - 1)}{q(q^2 - 1)} \cdot \frac{q(q^2 - 1)}{6} = \frac{q^n (q^{2n} - 1)}{6}.$$

This implies that each 3-subset of $PG(1, \mathbb{F}_{q^n})$ is contained in exactly one block of B^G. Hence, the pair $(\mathcal{M}, B^G)$ is a $3 - (q^n + 1, q + 1, 1)$ design. This design is called a Möbius plane, or an inverse plane.

6.4   Comments

In this chapter some well-known constructions of 2-designs and 3-designs are described. The construction for $STS(v)$ with $v \equiv 3 \ (mod \ 6)$ is due to Bose [3], and that for the case of $v \equiv 1 \ (mod \ 6)$ is due to Skolem [44]. Both were found during the twentieth century.

The Steiner Quadruple System was found by Kirkman [18], and the Möbius plane was found by Witt [58] in 1938.

6.5   Exercises

6.1 Construct a $STS(7)$ by the Skolem construction.

6.2 Construct a $STS(15)$ by the Bose construction.

6.3 Let $A = (x_1, y_1)$ and $B = (x_2, y_2)$ be two distinct points of $AG(2, \mathbb{F}_q)$. Find the equation which defines the line passing through these two points in $AG(2, \mathbb{F}_q)$.

6.4 Construct a $2 - (7, 3, 1)$ design by using the projective plane $PG(2, \mathbb{F}_2)$.

6.5 Construct a $2 - (9, 3, 1)$ design by using the affine plane $AG(2, \mathbb{F}_3)$.

6.6 Construct a $2 - (q^n, q, 1)$ and a $2 - (q^n + q^{n-1} + ... + q + 1, q + 1, 1)$ design using the spaces $AG(n, \mathbb{F}_q)$ and $PG(n, \mathbb{F}_q)$, respectively.

Chapter 7

Orthogonal Arrays of Index Unity

We have shown that orthogonal arrays (**OA**) of index unity can be used to construct perfect Cartesian codes (Theorem 3.2). Several known constructions of orthogonal arrays of index unity will be described in this chapter. Section 7.1 presents the construction of orthogonal arrays with strength 2 by using an affine plane over finite fields and orthogonal Latin squares. Sections 7.2 and 7.3 are devoted to the proof of the existence of $OA(n^2, 4, n, 2)$ for any positive integers n except $n = 2, 6$, which is equivalent to the existence of orthogonal Latin squares of order $n \neq 2, 6$. Section 7.4 describes the constructions due to Bush [7]. Section 7.5 explores the connections between orthogonal arrays and error-correcting codes. It is shown in Section 7.6 that all maximal distance separable (MDS) codes are equivalent to orthogonal arrays of index unity. The construction using Reed–Solomon codes, which are MDS codes, is described in this section.

Let $OA(N, k, s, t)$ denote an orthogonal array with N rows, k columns, s levels, and strength t.

7.1 OA with Strength $t = 2$ and Orthogonal Latin Squares

Let $AG(2, \mathbb{F}_q)$ be the affine plane over the finite field $\mathbb{F}_q$. Any two lines on $AG(2, \mathbb{F}_q)$,

$$ax + by + c = 0,$$

$$ax + by + c' = 0,$$

where $a, b, c, c' \in \mathbb{F}_q$ and $c \neq c'$, are said to be parallel. Given a fixed pair (a, b) we may have q parallel lines when the element c runs through $\mathbb{F}_q$. For any given point there is a unique line in this group of parallel lines which passes through the point. When the pair (a, b) runs through the set

$$\{(1, \alpha) | \alpha \in \mathbb{F}_q\} \cup \{0, 1\}$$

we get $q + 1$ groups of parallel lines which are denoted by $\phi_0, \phi_1, ..., \phi_q$. We label the q lines in each group by the numbers $1, 2, ..., q$.

Construct a $q^2 \times (q+1)$ array A whose columns are labeled with the groups $\phi_i (0 \leq i \leq q)$ and whose rows are labeled with the points of $AG(2, \mathbb{F}_q) = \{(\alpha, \beta) \mid \alpha, \beta \in \mathbb{F}_q\}$. The entry in this array in the row labeled (α, β) and the column labeled ϕ_j is defined to be the labeled number of the line passing through (α, β) in ϕ_j. Taking one line from each of any two groups ϕ_i, ϕ_j, these two lines have a unique common point since they are not parallel. This proves that the array A is an $OA(q^2, q+1, q, 2)$.

The orthogonal arrays with strength $t = 2$ have a close relation with orthogonal Latin squares. A Latin square of order n is an $n \times n$ array with entries from a set S of cardinality n such that each element of S appears once in every row and every column. It is easily seen that a Latin square of order n exists for every positive n; the following is an example:

0	1	2	...	$n-1$
1	2	3	...	0
...				
$n-1$	0	1	...	$n-2$

Two Latin squares of order n are said to be orthogonal to each other if, when one is superimposed on the other, the ordered pairs (i, j) of corresponding entries consist of all possible n^2 pairs. A collection of w Latin squares of order n, any pair of which are orthogonal, is called a set of Mutually Orthogonal Latin Squares (*MOLS*) of order n, and denoted by $MOLS(n, w)$. Let $N(n)$ denote the maximum number of *MOLS* of order n.

PROPOSITION 7.1
There do not exist $MOLS(n, n)$ if $n > 1$ {i.e., $N(n) \leq n - 1$ for $n > 1$}.

PROOF   Suppose that $L_1, \cdots, L_w$ are mutually orthogonal Latin squares of order n. Without loss of generality, we may assume that $L_1, \cdots, L_w$ are all defined on the symbol set $\{1, \cdots, n\}$. Furthermore, we may assume that the first row of each of these squares is

$$(1, 2, \cdots, n).$$

(This is justified by observing that within any L_i we can relabel the symbols so the first row is as specified. The relabeling does not affect the orthogonality of the squares.)

Now consider the w values $L_1(2, 1), \cdots, L_w(2, 1)$ (this is where we require the assumption $n \geq 2$). We first note that these w values are all distinct, as follows: suppose that $L_i(2, 1) = L_j(2, 1) = x$. Then we have the ordered pair (x, x) occurring in the superposition of L_i and L_j in cell $(1, x)$ and again in cell $(2, 1)$. This contradicts the orthogonality of L_i and L_j.

Next we observe that $L_i(2, 1) \neq 1$ for $1 \leq i \leq w$. This follows from the fact that $L_i(1, 1) = 1$ and no symbol can occur in two cells in any column of a Latin square.

Combining our two observations, we see that $L_1(2, 1), \cdots, L_w(2, 1)$ are in fact w distinct elements from the set $\{2, \cdots, n\}$. Hence, $w \leq n - 1$. $\quad$ ▯

THEOREM 7.1

If $q > 2$ is a prime power then there are $q - 1$ mutually orthogonal Latin squares of order q.

PROOF Let the elements of $\mathbb{F}_q$ be denoted by $\alpha_0 = 0, \alpha_1, ..., \alpha_{q-1}$. For $i = 1, 2, ..., q - 1$ define a Latin square L_i by taking the entry in row j and column n to be $\alpha_i \alpha_j + \alpha_k$, for $j, k = 0, 1, ..., q - 1$.

For fixed $i \in \{1, 2, ..., q - 1\}$ and $j \in \{0, 1, ..., q - 1\}$ it is clear that the elements

$$\alpha_i \alpha_j + \alpha_k, \quad k = 0, 1, ..., q - 1$$

comprise all of $\mathbb{F}_q$. Similarly, if $i \in \{1, 2, ..., q - 1\}$ and $k \in \{0, 1, ..., q - 1\}$ are fixed, it is clear that

$$\alpha_i \alpha_j + \alpha_k, \quad j = 0, 1, ..., q - 1$$

comprise all of $\mathbb{F}_q$. This shows that each L_i is indeed a Latin square.

Fix $i, i' \in \{1, 2, ..., q - 1\}, i \neq i'$. For any $z_1, z_2 \in \mathbb{F}_q$ consider the linear equations

$$\alpha_i \alpha_j + \alpha_k = z_1,$$

$$\alpha_{i'} \alpha_j + \alpha_k = z_2.$$

In the unknowns α_j and α_k, there is a unique solution (α'_j, α'_k). Consequently, any ordered pair (z_1, z_2) appears exactly once when L_i is superimposed on $L_{i'}$, which establishes the orthogonality of the Latin squares. $\quad$ ▯

Example 7.1

Taking $L_1(j, k) \equiv j + k \pmod 3$, $L_2(j, k) \equiv 2j + k \pmod 3$, we obtain the orthogonal Latin squares of order 3:

$$L_1 = \begin{pmatrix} 0 & 1 & 2 \\ 1 & 2 & 0 \\ 2 & 0 & 1 \end{pmatrix}, \quad L_2 = \begin{pmatrix} 0 & 1 & 2 \\ 2 & 0 & 1 \\ 1 & 2 & 0 \end{pmatrix}.$$

▯

Example 7.2

Taking $\mathbb{F}_4 = \{0, 1, \alpha, 1 + \alpha\}$ where $\alpha^2 + \alpha + 1 = 0$, let $\alpha_0 = 0, \alpha_1 = 1, \alpha_2 = \alpha, \alpha_3 = 1 + \alpha$ and $\alpha_{L_i(j,k)} = \alpha_i \alpha_j + \alpha_k$ where $i = 1, 2, 3, 0 \leq j, k \leq 3$. We obtain a $MOLS(4,3)$:

$$L_1 = \begin{pmatrix} 0\ 1\ 2\ 3 \\ 1\ 0\ 3\ 2 \\ 2\ 3\ 0\ 1 \\ 3\ 2\ 1\ 0 \end{pmatrix}, \quad L_2 = \begin{pmatrix} 0\ 1\ 2\ 3 \\ 2\ 3\ 0\ 1 \\ 1\ 0\ 3\ 2 \\ 3\ 2\ 1\ 0 \end{pmatrix}, \quad L_3 = \begin{pmatrix} 0\ 1\ 2\ 3 \\ 3\ 2\ 1\ 0 \\ 2\ 3\ 0\ 1 \\ 1\ 0\ 3\ 2 \end{pmatrix}.$$

THEOREM 7.2
A set of $MOLS(n,w)$ is equivalent to an $OA(n^2, w+2, n, 2)$.

PROOF   Let these w Latin squares named by $L_1, \cdots, L_w$ be defined on the symbol set $\{1, \cdots, n\}$, and have rows and columns labeled by $\{1, \cdots, n\}$. Convert each of the w Latin squares to an $n^2 \times 1$ array by juxtaposing the s rows of the square and transposing. Combine these arrays to form an $n^2 \times w$ array. Add two more columns to this array such that each row is a $(w+2)$-tuple:

$$(i, j, L_1(i,j), \cdots, L_w(i,j)).$$

These form an $n^2 \times (w+2)$ array A. We will prove that A is an $OA(n^2, w+2, n, 2)$.

We need to show that every ordered pair of symbols occurs in any two columns a and b, where $1 \le a < b \le w+2$. We consider in several cases:

1. If $a = 1$ and $b = 2$, then clearly we get every ordered pair.

2. If $a = 1$ and $b \ge 3$, then we get every ordered pair because every row of L_b is a permutation of $\{1, \cdots, n\}$.

3. If $a = 2$ and $b \ge 3$, then we get every ordered pair because every column of L_b is a permutation of $\{1, \cdots, n\}$.

4. If $a \ge 3$, then we get every ordered pair because L_a and L_b are orthogonal.

The construction can easily be reversed. If A is an $OA(n^2, k, n, 2)$ with $k \ge 3$, then we can construct a set of $MOLS(n, k-2)$ from it. Suppose that A is defined on the symbol set $\{1, \ldots, n\}$. Label the columns of A by the integers $1, \cdots, k$, and label the rows of A by the integers $1, \cdots, n^2$. We construct $MOLS(n, k-2)$, which named $L_1, \cdots, L_{k-2}$, as follows. For $1 \le h \le k-2$ and $1 \le r \le n^2$, define

$$L_h(A(r,1), A(r,2)) = A(r, h+2).$$

We will show that $L_1, \cdots, L_{k-2}$ are orthogonal Latin squares of order n.

We will show that each L_h is a Latin square first. Each cell contains one and only one entry because every ordered pair occurs exactly once in columns 1 and 2 of A. Next, let us show that each row i of L_h is a permutation of $\{1, \cdots, n\}$. The entries in row i of L_h in fact are the symbols in the set

$$\{A(r, h + 2) \ : \ A(r, 1) = i\}.$$

These symbols are all distinct because every ordered pair occurs exactly in columns 1 and $h + 2$ of A. A similar argument proves that each column j of each L_h is a permutation of $\{1, \cdots, n\}$. Hence, $L_1, \cdots, L_{k-2}$ are all Latin squares.

It remains to prove orthogonality. But L_h and L_g are orthogonal because every ordered pair occurs exactly once in columns $h + 2$ and $g + 2$ of A. ☐

Combining Theorems 7.1 and 7.2 we can construct an $OA(q^2, q + 1, q, 2)$ when q is a prime power.

Example 7.3
The following $OA(9, 4, 3, 2)$ can be constructed from the orthogonal Latin squares of order 3 given in Example 7.1:

$$\begin{pmatrix} 0\ 0\ 0\ 0 \\ 0\ 1\ 1\ 1 \\ 0\ 2\ 2\ 2 \\ 1\ 0\ 1\ 2 \\ 1\ 1\ 2\ 0 \\ 1\ 2\ 0\ 1 \\ 2\ 0\ 2\ 1 \\ 2\ 1\ 0\ 2 \\ 2\ 2\ 1\ 0 \end{pmatrix}$$

☐

We will look at several constructions of orthogonal Latin squares. First we give a construction that works for all odd integers.

THEOREM 7.3
If $n > 1$ is odd, then there exist orthogonal Latin squares of order n.

PROOF   We define two Latin squares of order n with entries from $\mathbb{Z}_n$:

$$L_1(i, j) \equiv i + j \ (mod \ n)$$
$$L_2(i, j) \equiv i - j \ (mod \ n)$$

L_1 and L_2 are easily seen to be Latin squares for any positive integer n. Let us prove that they are orthogonal when n is odd. Suppose that $(x, y) \in \mathbb{Z}_n \times \mathbb{Z}_n$. We want to find a unique cell (i, j) such that $L_1(i, j) = x$ and $L_2(i, j) = y$. In other words, we want to solve the system of congruence equations:

$$i + j \equiv x \ (mod \ n)$$
$$i - j \equiv y \ (mod \ n)$$

for i and j. Since n is odd, 2 has a multiplicative inverse modulo n, and the system has the unique solution

$$i \equiv (x + y)2^{-1} \ (mod \ n)$$
$$j \equiv (x - y)2^{-1} \ (mod \ n).$$

Hence L_1 and L_2 are orthogonal. ☐

We construct orthogonal Latin squares of order 5 using Theorem 7.3:

$$L_1 = \begin{pmatrix} 0\ 1\ 2\ 3\ 4 \\ 1\ 2\ 3\ 4\ 0 \\ 2\ 3\ 4\ 0\ 1 \\ 3\ 4\ 0\ 1\ 2 \\ 4\ 0\ 1\ 2\ 3 \end{pmatrix}, \quad L_2 = \begin{pmatrix} 0\ 4\ 3\ 2\ 1 \\ 1\ 0\ 4\ 3\ 2 \\ 2\ 1\ 0\ 4\ 3 \\ 3\ 2\ 1\ 0\ 4 \\ 4\ 3\ 2\ 1\ 0 \end{pmatrix}$$

Suppose that L and M are Latin squares of order m and n defined on the symbol sets X and Y, respectively. We define the direct product of L and M, denoted by $L \times M$, to be the $mn \times mn$ array defined by

$$(L \times M)((i_1, i_2), (j_1, j_2)) = (L(i_1, j_1), M(i_2, j_2)).$$

LEMMA 7.1
If L and M are Latin squares of order m and n defined on symbol sets X and Y, respectively, then $L \times M$ is a Latin square of order mn defined on symbol set $X \times Y$.

PROOF   Consider a row, say (i_1, i_2), of $L \times M$. Let $x \in X$ and $y \in Y$. We will show how to find the symbol (x, y) in row (i_1, i_2). Since L is a Latin square, there is a unique column j_1 such that $L(i_1, j_1) = x$. Since M is a Latin square, there is a unique column j_2 such that $M(i_2, j_2) = y$. Then $(L \times M)\{(i_1, i_2), (j_1, j_2)\} = (x, y)$.

Similarly, every column of $L \times M$ contains every symbol in $X \times Y$, so $L \times M$ is a Latin square. ☐

Example 7.4
The following is an example of the direct product. Suppose L and M are as follows:

$$L = \begin{pmatrix} 3\ 1\ 2 \\ 2\ 3\ 1 \\ 1\ 2\ 3 \end{pmatrix}, \quad M = \begin{pmatrix} 1\ 2 \\ 2\ 1 \end{pmatrix}$$

Then $L \times M$ is as follows:

$$\begin{pmatrix}
(3,1)\ (1,1)\ (2,1)\ (3,2)\ (1,2)\ (2,2) \\
(2,1)\ (3,1)\ (1,1)\ (2,2)\ (3,2)\ (1,2) \\
(1,1)\ (2,1)\ (3,1)\ (1,2)\ (2,2)\ (3,2) \\
(3,2)\ (1,2)\ (2,2)\ (3,1)\ (1,1)\ (2,1) \\
(2,2)\ (3,2)\ (1,2)\ (2,1)\ (3,1)\ (1,1) \\
(1,2)\ (2,2)\ (3,2)\ (1,1)\ (2,1)\ (3,1)
\end{pmatrix}$$

The direct product $L \times M$ contains many copies of L and M within it. The Latin square $L \times M$ can be partitioned into m^2 disjoint $n \times n$ subarrays, each of which is a copy of M on the symbol set $\{x\} \times Y$, where $x \in X$. $L \times M$ can also be partitioned into n^2 disjoint $m \times m$ subarrays, each of which is a copy of L on the symbol set $X \times \{y\}$, where $y \in Y$.

Next we prove that the direct product construction preserves orthogonality.

PROPOSITION 7.2
If there exist orthogonal Latin squares of orders n_1 and n_2, then there exist orthogonal Latin squares of order $n_1 n_2$.

PROOF Suppose that L_1 and L_2 are orthogonal Latin squares of order n_1 on symbol set X, and M_1 and M_2 are orthogonal Latin squares of order n_2 on symbol set Y. We will show that $L_1 \times M_1$ and $L_2 \times M_2$ are orthogonal Latin squares of order $n_1 n_2$. $L_1 \times M_1$ and $L_2 \times M_2$ are both Latin squares by Lemma 7.1, so we just have to prove that they are orthogonal.

Consider the ordered pair of symbols $\{(x_1, y_1), (x_2, y_2)\}$. We want to find a unique cell $\{(i_1, i_2), (j_1, j_2)\}$ such that

$$(L_1 \times M_1)\{(i_1, i_2), (j_1, j_2)\} = (x_1, y_1)$$
$$(L_2 \times M_2)\{(i_1, i_2), (j_1, j_2)\} = (x_2, y_2).$$

This is equivalent to

$$L_1(i_1, j_1) = x_1,$$
$$M_1(i_2, j_2) = y_1,$$
$$L_2(i_1, j_1) = x_2,$$
$$M_2(i_2, j_2) = y_2$$

The first and third equations determine (i_1, j_1) uniquely because L_1 and L_2 are orthogonal; and the second and fourth equations determine (i_2, j_2) uniquely because M_1 and M_2 are orthogonal. The desired cell $\{(i_1, i_2), (j_1, j_2)\}$ is determined uniquely. ▯

The following proposition can be proved by Theorems 7.1, 7.2, and 7.3 with Proposition 7.2.

PROPOSITION 7.3
There exist orthogonal Latin squares of order n if $n \not\equiv 2 \pmod 4$.

PROOF   If n is odd, then apply Theorem 7.3. If n is a power of 2 and $n \geq 4$, then apply Theorem 7.1. Finally, suppose that n is even, $n \not\equiv 2 \pmod 4$, and n is not a power of 2. Then we can write $n = 2^i n'$, where $i \geq 2$ and $n' > 1$ is odd. Applying Proposition 7.2, the result follows. ▯

The direct product construction of Proposition 7.2 can be generalized to the direct product of $MOLS(n_1, s)$ and $MOLS(n_2, s)$, where $s > 2$, in an obvious way. It is also possible to form the direct product of $MOLS(n_i, s)$, $1 \leq i \leq k$, where $k > 2$. Thus we have the following proposition.

PROPOSITION 7.4
If there exist $MOLS(n_i, s)$, $1 \leq i \leq k$, then there exists $MOLS(n, s)$ where $n = n_1 \times n_2 \times \cdots \times n_k$.

Based on Theorem 7.1 and Proposition 7.4 we have the following proposition.

PROPOSITION 7.5
Suppose that n has prime factorization $n = p_1^{e_1} \cdots p_k^{e_k}$, where the $p_i's$ are distinct primes and $e_i \geq 1$ for $1 \leq i \leq k$. Let

$$s = \min\{p_i^{e_i} - 1 \; : \; 1 \leq i \leq k\}.$$

Then there exists $MOLS(n, s)$.

There are many corollaries of Proposition 7.5 that can be proven. Here is a specific result that we will use later.

COROLLARY 7.1
If $n \equiv 1, 5, 7$, or $11 \pmod{12}$, then there exists $MOLS(n, 4)$. If $n \equiv 4$ or 8 $\pmod{12}$, there exists $MOLS(n, 3)$.

PROOF Suppose that n has prime power factorization $n = p_1^{e_1} \cdots p_k^{e_k}$. By Proposition 7.5, there exists a $MOLS(n, 3)$ if $p_i^{e_i} \geq 4$ for $1 \leq i \leq k$. The only situations in which $p_i^{e_i} < 4$ are when $(p_i, e_i) = (2, 1)$ or $(p_i, e_i) = (3, 1)$. In other words, if the prime power factorization of n does not contain the specific terms 2 or 3, then a $MOLS(n, 3)$ exists. By a similar argument, if the prime power factorization of n does not contain the specific terms $2, 2^2$ or 3, then a $MOLS(n, 4)$ exists.

Now, if $n \equiv 1, 5, 7, \text{ or } 11 \ (mod \ 12)$, then $gcd(n, 6) = 1$, so there are no terms involving 2 or 3 in the factorization of n. It follows that a $MOLS(n, 4)$ exists.

If $n \equiv 4, 8 \ (mod \ 12)$, then 4 is a factor of n and 3 is not a factor of n. Therefore there is no term involving 3 in the factorization of n, and the term involving 2 has an exponent that is at least 2. Hence a $MOLS(n, 3)$ exists. $\square$

7.2   Transversal Designs

A transversal design is another type of design equivalent to $OA(n^2, k, n, 2)$.

DEFINITION 7.1   *Let $k \geq 2$ and $n \geq 1$. A transversal design $TD(k, n)$ is a triple $(\mathcal{X}, \mathcal{G}, \mathcal{B})$ such that the following properties are satisfied:*

1. *$\mathcal{X}$ is a set of kn elements called points,*

2. *$\mathcal{G}$ is a partition of $\mathcal{X}$ into k subsets of size n called groups,*

3. *$\mathcal{B}$ is a set of k-subsets of $\mathcal{X}$ called blocks,*

4. *any group and any block contain exactly one common point,*

5. *every pair of points from distinct groups is contained in exactly one block.*

It is easy to find that each point is contained in n blocks and there are totally n^2 blocks in a $TD(k, n)$.

We first show how to construct a $TD(k, n)$ from an $OA(n^2, k, n, 2)$. Let A be an $OA(n^2, k, n, 2)$ on the symbol set $\{1, \cdots, n\}$. Label the columns of A as $1, \cdots, k$, and label the rows of A as $1, \cdots, n^2$. Define

$$\mathcal{X} = \{(i, j) \mid 1 \leq i \leq n, \ 1 \leq j \leq k\}.$$

For $1 \leq j \leq k$, define

$$G_j = \{(i, j) \mid 1 \leq i \leq n\},$$

and then define

$$\mathscr{G} = \{G_j \; : \; 1 \leq j \leq k\}.$$

For $1 \leq r \leq n^2$, define

$$B_r = \{(A(r,j),j) \; : \; 1 \leq j \leq k\},$$

and define

$$\mathscr{B} = \{B_r \; : \; 1 \leq r \leq n^2\}.$$

It is trivial to prove that $(\mathscr{X}, \mathscr{G}, \mathscr{B})$ is a $TD(k,n)$.

Example 7.5

From the $OA(9,4,3,2)$ constructed in Example 7.3 we obtain a $TD(k,n)$ with $k = 4, n = 3$. Its blocks are shown in the following:

$$B_1 = \{(1,1),(1,2),(1,3),(1,4)\}$$
$$B_2 = \{(1,1),(2,2),(2,3),(2,4)\}$$
$$B_3 = \{(1,1),(3,2),(3,3),(3,4)\}$$
$$B_4 = \{(2,1),(1,2),(2,3),(3,4)\}$$
$$B_5 = \{(2,1),(2,2),(3,3),(1,4)\}$$
$$B_6 = \{(2,1),(3,2),(1,3),(2,4)\}$$
$$B_7 = \{(3,1),(1,2),(3,3),(2,4)\}$$
$$B_8 = \{(3,1),(2,2),(1,3),(3,4)\}$$
$$B_9 = \{(3,1),(3,2),(2,3),(1,4)\}$$

❑

The construction can be reversed: given a $TD(k,n)$, we can use it to construct an $OA(n^2,k,n,2)$. Suppose $(\mathscr{X}, \mathscr{G}, \mathscr{B})$ is a $TD(k,n)$. We can assume that $\mathscr{X} = \{(i,j) \mid 1 \leq i \leq n, \ 1 \leq j \leq k\}$ and $\mathscr{G} = \{G_j \; : \; 1 \leq j \leq k\}$, where $G_j = \{(i,j) \mid 1 \leq i \leq n\}$ for $1 \leq j \leq k$. For each block $B \in \mathscr{B}$ and for $1 \leq j \leq k$, let $(b_j,j) \in B \cap G_j$ (recall that each block intersects each group in a unique point). Then, for each $B \in \mathscr{B}$, form the k-tuple

$$(b_1, \cdots, b_k).$$

Constructing an array A whose rows consist of all these k-tuples, it is easy to see that A is an $OA(n^2,k,n,2)$. Thus, we have proved the following proposition:

PROPOSITION 7.6

Suppose that $n \geq 2$ and $k \geq 3$. Then the existence of any one of the following designs implies the existence of the other two designs:

1. *a MOLS$(n, k-2)$,*

2. *an OA$(n^2, k, n, 2)$,*

3. *a TD(k, n).*

Next we introduce the direct product of two transversal designs which is a generalization of the direct product for two Latin squares. A $TD(k, tm)$ can be constructed from a $TD(k, t)$ and a $TD(k, m)$ by the following method.

Let $(\mathscr{X}, \mathscr{G}, \mathscr{A})$ be a $TD(k, t)$, where $G_1, \cdots, G_k$ are the groups. Design

$$\mathscr{Y} = \mathscr{X} \times \{1, \cdots, m\},$$

and for $1 \leq j \leq k$ define

$$H_j = G_j \times \{1, \cdots, m\}.$$

Let $\mathscr{H} = \{H_j \; : \; 1 \leq j \leq k\}$. $\mathscr{Y}$ and $\mathscr{H}$ will be the points and groups, respectively, of the $TD(k, tm)$ that we are constructing.

We now define the blocks of this transversal design. For every $A \in \mathscr{A}$, construct m^2 blocks as follows. For $1 \leq j \leq k$, let $\{a_j\} = A \cap G_j$. Then let $\mathscr{B}_A$ be the set of m^2 blocks of a $TD(k, m)$ in which the groups are

$$\{a_j\} \times \{1, \cdots, m\}, \quad 1 \leq j \leq k.$$

Then define

$$\mathscr{B} = \bigcup_{A \in \mathscr{A}} \mathscr{B}_A.$$

We show that $(\mathscr{Y}, \mathscr{H}, \mathscr{B})$ is a transversal design. The main task is to show that any two points from different groups occur in a unique block. Suppose that $x = (g, a)$ and $y = (h, b)$, where $g \in G_i$, $h \in G_j$, $i \neq j$, and $a, b \in \{1, \cdots, m\}$. There is a unique block $A \in \mathscr{A}$ such that $g, h \in A$ because g and h occur in different groups in $\mathscr{G}$. Then it is easily seen that x and y occur in a unique block in $\mathscr{B}_A$ and in no other block in $\mathscr{B}$.

We now describe the Wilson's construction for $MOLS$. This construction uses a type of design called a truncated transversal design, which is formed from a transversal design by deleting some points from one of the groups. More specifically, let $(\mathscr{X}, \mathscr{G}, \mathscr{B})$ be a $TD(k+1, t)$, where $k \geq 2$. Pick a group $G \in \mathscr{G}$, and suppose that $1 \leq u \leq t$. Let $G' \subseteq G$, $|G'| = u$. Then define

$$\mathscr{Y} = (\mathscr{X} \setminus G) \cup G'$$
$$\mathscr{H} = (\mathscr{G} \setminus \{G\}) \cup \{G'\}$$
$$\mathscr{C} = \{B \in \mathscr{B} : B \cap G' \neq \emptyset\} \cup \{B \setminus \{x\} : B \in \mathscr{B}, B \cap G = \{x\}, x \in G \setminus G'\}$$

The set system $(\mathscr{Y}, \mathscr{H}, \mathscr{C})$ is a truncated transversal design. If $u < t$, then this design has $kt + u$ points, k groups of size t and one group of size u, $t(t-u)$

blocks of size k, and tu blocks of size $k + 1$. {If $t = u$, then the design is just a $TD(k + 1, t)$ because we have deleted no points.}

Wilson's Construction

THEOREM 7.4
Let $k \geq 2$ and suppose that the following transversal designs exist: a $TD(k + 1, t)$, a $TD(k, m)$, a $TD(k, m + 1)$, and a $TD(k, u)$, where $1 \leq u \leq t$. Then there exists a $TD(k, mt + u)$.

PROOF   First construct a truncated transversal design from a $TD(k+1, t)$ by deleting $t - u$ points from some group, as described above. Let $(\mathcal{X}, \mathcal{G}, \mathcal{A})$ be the resulting truncated transversal design, where $G_1, \cdots, G_k$ are k groups of size t and G_{k+1} is a group of size u.
Define

$$\mathcal{Y} = ((\mathcal{X} \setminus G_{k+1}) \times \{1, \cdots, m\}) \cup (\{1, \cdots, k\} \times G_{k+1}).$$

Then for $1 \leq j \leq k$, define

$$H_j = (G_j \times \{1, \cdots, m\}) \cup (\{j\} \times G_{k+1}),$$

and let $\mathcal{H} = \{H_j : 1 \leq j \leq k\}$. $\mathcal{Y}$ and $\mathcal{H}$ will be the points and groups, respectively, of the $TD(k, mt + u)$ that we are constructing.
It will be convenient to define a "type I" point to be a point in $(\mathcal{X} \setminus G_{k+1}) \times \{1, \cdots, m\}$, and a "type II" point to be a point in $\{1, \cdots, k\} \times G_{k+1}$. Observe that each group H_j contains mt type I points (which consist of m copies of each point in G_j) and u type II points (which consist of one copy of each point in G_{k+1}).
We now define the blocks of this transversal design. For each block $A \in \mathcal{A}$, construct a set of blocks $\mathcal{B}_A$ according to the following recipe:

1. Suppose $|A| = k$. For $1 \leq j \leq k$, let $\{a_j\} = A \cap G_j$. Then let $\mathcal{B}_A$ be the set of m^2 blocks of a $TD(k, m)$ in which the groups are

$$\{a_j\} \times \{1, \cdots, m\}, \quad 1 \leq j \leq k.$$

   Observe that the blocks in $\mathcal{B}_A$ contain only type I points.

2. Suppose $|A| = k + 1$. For $1 \leq j \leq k + 1$, let $\{a_j\} = A \cap G_j$. There exists a $TD(k, m + 1)$ in which the groups are

$$(\{a_j\} \times \{1, \cdots, m\} \cup \{(j, a_{k+1})\}, \quad 1 \leq j \leq k,$$

   and in which

$$\{(1, a_{k+1}), \cdots, (k, a_{k+1})\}$$

is a block, which is the only one block containing more than one type II point. Delete this block, and let $\mathscr{B}_A$ be the set of $(m+1)^2 - 1$ blocks that remain. No block in $\mathscr{B}_A$ contains more than one type II point.

Finally, there exists a $TD(k, u)$ in which the groups are

$$\{j\} \times G_{k+1} \qquad 1 \leq j \leq k.$$

Let $\mathscr{B}*$ denote the blocks of this transversal design. The blocks in $\mathscr{B}*$ contain only type II points.

The block set of the $TD(k, tm + u)$ is defined to be

$$\mathscr{B} = \left(\bigcup_{A \in \mathscr{A}} \mathscr{B}_A \right) \cup \mathscr{B}*.$$

The number of blocks in $\mathscr{B}$ is

$$(t^2 - tu)m^2 + tu((m+1)^2 - 1) + u^2 = (tm + u)^2.$$

In order to prove that $(\mathscr{Y}, \mathscr{H}, \mathscr{B})$ is a transversal design, we need to show that any two points x and y from different groups of $\mathscr{H}$ occur in a unique block. There are three different cases to consider according to the types of the two points x and y.

1. Suppose x and y are both of type I. Let $X = (g, a)$ and $y = (h, b)$, where $g \in G_i$, $h \in G_j$, $i \neq j$, and $a, b \in \{1, \cdots, m\}$. There is a unique block $A \in \mathscr{A}$ such that $g, h \in A$. Then x and y occur in a unique block in $\mathscr{B}_A$ and in no other block in $\mathscr{B}$.

2. Suppose x is of type I and y is of type II. Let $x = (g, a)$ and $y = (j, h)$, where $g \in G_i$, $h \in G_{k+1}$, $a \in \{1, \cdots, m\}$, and $j \in \{1, \cdots, k\}$. There is a unique block A such that $g, h \in A$, and it must be the case that $|A| = k + 1$. x and y occur in a unique block in $\mathscr{B}_A$ and in no other block in $\mathscr{B}$.

3. Suppose x and y are both of type II. Let $x = (i, g)$ and $y = (j, h)$, where $g, h \in G_{k+1}$, $i, j \in \{1, \cdots, k\}$, and $i \neq j$. Then x and y occur in a unique block in $\mathscr{B}*$ and in no other blocks in $\mathscr{B}$ (note that the blocks in $\mathscr{B}*$ are the only ones that contain more than one point of type II).

This completes the proof. $\qquad\qquad\qquad\qquad\qquad\qquad\qquad\qquad$ ꕔ

In view of Proposition 7.6, the following corollary is obtained from Theorem 7.4 in the language of MOLS.

COROLLARY 7.2

Suppose $s \geq 1$ and there exist a $MOLS(t, s+1)$, a $MOLS(m, s)$, a $MOLS(m+1, s)$, and a $MOLS(u, s)$ where $1 \leq u \leq t$. Then there exists a $MOLS(mt + u, s)$.

7.3 Existence of $OA(n^2, 4, n, 2)$

Some orthogonal arrays with $t = 2$ have been constructed in Section 7.1 when the level n is a prime power. Now we consider the case when the level n is a general positive integer. We will prove that if $n \neq 2$ or 6, then there exists an $OA(n^2, 4, n, 2)$, which is equivalent to the existence of orthogonal Latin squares of order n by Theorem 7.2.

PROPOSITION 7.7
Suppose $t \equiv 1, 5 \ (mod \ 6)$, u is odd, and $1 \leq u \leq t$. Then there exist orthogonal Latin squares of order $3t + u$.

PROOF   We apply Corollary 7.2 with $s = 2$ and $m = 3$, noting that orthogonal Latin squares of order 3, 4 and u exist, as does a $MOLS(t, 3)$ (Corollary 7.1).   ◻

PROPOSITION 7.8
For all positive integers $n \equiv 10 \ (mod \ 12)$, there exists an $OA(n^2, 4, n, 2)$ and hence there exist orthogonal Latin squares of order n.

PROOF   Any positive integer $n \equiv 10 \ (mod \ 12)$ can be written in the form $n = 3m + 1$, where $m \equiv 3 \ (mod \ 4)$. Define $X = \mathbb{Z}_{2m+1} \cup \Omega$, where $\Omega = \{\infty_i : 1 \leq i \leq m\}$. Begin with the following $4m + 1$ 4-tuples:

$$(0, 0, 0, 0),$$
$$(0, 2i, i, \infty_i), \qquad 1 \leq i \leq m,$$
$$(0, 2i - 1, i, \infty_i), \qquad 1 \leq i \leq m,$$
$$(0, \infty_i, 2m + 1 - i, i), \qquad 1 \leq i \leq m,$$
$$(\infty_i, 0, i, 2m + 1 - i), \qquad 1 \leq i \leq m.$$

Next develop each of these $4m + 1$ 4-tuples through the group $\mathbb{Z}_{2m+1}$ using the convention that $\infty_i + j = \infty_i$ for all $j \in \mathbb{Z}_{2m+1}$ and all $1 \leq i \leq m$. Call the resulting set of $(4m + 1)(2m + 1)$ 4-tuples A_1.

Now let A_2 be an $OA(m^2, 4, m, 2)$ on the symbol set Ω. {Note that if m is odd, then there exist orthogonal Latin squares of order m, therefore, there exists an $OA(m^2, 4, m, 2)$.} A_2 contains m^2 4-tuples.

The $(4m + 1)(2m + 1) + m^2 = (3m + 1)^2$ 4-tuples in $A_1 \cup A_2$ form an $OA([3m + 1]^2, 4, 3m + 1, 2)$ that we are constructing. This orthogonal array has the following permutation α as an automorphism:

$$\alpha = (0, 1, 2, \cdots, 2m)(\infty_1) \cdots (\infty_m).$$

The $(3m + 1)^2$ 4-tuples in this array are comprised of $4m + 1$ orbits each consisting of $2m + 1$ rows and m^2 orbits each consisting of one row. Note that the orbit representatives of all ordered pairs of symbols from X under the group $G = \{\alpha^i : 0 \le i \le 2m\}$ are as follows:

$$
\begin{aligned}
(0, i), && 0 \le i \le 2m \\
(0, \infty_i), && 1 \le i \le m \\
(\infty_i, 0), && 1 \le i \le m \\
(\infty_i, \infty_j), && 1 \le i, j \le m.
\end{aligned}
$$

It is not hard to show that for each choice of two columns, every orbit of ordered pairs is contained in exactly one of the orbits of 4-tuples, within the specified columns. Thus, we have constructed an $OA(n^2, 4, n, 2)$ as desired. □

PROPOSITION 7.9

There exist orthogonal Latin squares of order 14.

PROOF We present a set of 17 4-tuples of elements in $\mathbb{Z}_{11} \cup \{\infty_1, \infty_2, \infty_3\}$:

$$
\begin{array}{cccc}
0 & 0 & 0 & 0 \\
0 & 4 & 1 & 6 \\
4 & 0 & 6 & 1 \\
6 & 1 & 0 & 4 \\
1 & 6 & 4 & 0 \\
\infty_1 & 4 & 0 & 1 \\
\infty_2 & 6 & 0 & 2 \\
\infty_3 & 9 & 0 & 8 \\
4 & \infty_1 & 1 & 0 \\
6 & \infty_2 & 2 & 0 \\
9 & \infty_3 & 8 & 0 \\
1 & 0 & \infty_1 & 4 \\
2 & 0 & \infty_2 & 6 \\
8 & 0 & \infty_3 & 9 \\
0 & 1 & 4 & \infty_1 \\
0 & 2 & 6 & \infty_2 \\
0 & 8 & 9 & \infty_3 \\
\end{array}
$$

These rows are to be developed modulo 11 using the convention that $\infty_i + j = \infty_i$, for $i = 1, 2, 3$ and $j \in \mathbb{Z}_{11}$. Then adjoin nine more rows that form an $OA(9, 4, 3, 2)$ on the symbols $\infty_1, \infty_2, \infty_3$. The result is an $OA(196, 4, 14, 2)$, which is equivalent to the desired orthogonal Latin squares. □

PROPOSITION 7.10

Suppose $n \equiv 2 \pmod 4$, $n \neq 2, 6$. Then there exist orthogonal Latin squares of order n.

PROOF   We have already proved the cases where $n \equiv 10 \pmod{12}$ in Proposition 7.8, so we can assume that $n \equiv 2, 6, 14, 18, 26$, or $30 \pmod{36}$. For each of these six residue classes modulo 36, we present a construction that is an application of Proposition 7.7 by writing n in the form $n = 3t + u$ in an appropriate manner:

$$36s + 2 = 3(12s - 1) + 5, \quad s \geq 1$$
$$36s + 6 = 3(12s + 1) + 3, \quad s \geq 1$$
$$36s + 14 = 3(12s + 1) + 11, \quad s \geq 1$$
$$36s + 18 = 3(12s + 5) + 3, \quad s \geq 0$$
$$36s + 26 = 3(12s + 7) + 5, \quad s \geq 0$$
$$36s + 30 = 3(12s + 7) + 9, \quad s \geq 1.$$

The only values of n not covered by the constructions above are $n = 2, 6, 14$, and 30. The first two values of n are exceptions, the case of $n = 14$ is done in Proposition 7.9, and $n = 30$ can be handled by the direct product construction because $30 = 3 \times 10$ and orthogonal Latin squares of order 3 and 10 exist. ∎

Finally, we obtain our main existence result immediately from Propositions 7.3 and 7.10.

THEOREM 7.5

Suppose that n is a positive integer and $n \neq 2$, or 6. Then there exist orthogonal Latin squares, which is equivalent to the existence of $OA(n^2, 4, n, 2)$.

7.4   Bush's Construction

We have constructed some orthogonal arrays of index unity with the strength $t = 2$. It is not difficult to construct orthogonal arrays of index unity with $t \geq 2$.

Take the levels $0, 1, ..., s - 1$ as the residue classes modulo s and construct an $s^t \times t$ array which has each of the s^t possible t-tuple one time as a row. We add the entries in each row and let the negative of this sum be the level for a new column. The result $s^t \times (t + 1)$ array is an $OA(s^t, t + 1, s, t)$, since any t entries in a row determine the remainder entry in this row. This array has the property that the levels in every row add to zero, so it is called a zero-sum array. As an example, the following $OA(8, 4, 2, 3)$ is zero-sum:

$$\begin{pmatrix} 0\,0\,0\,0 \\ 0\,0\,1\,1 \\ 0\,1\,0\,1 \\ 0\,1\,1\,0 \\ 1\,0\,0\,1 \\ 1\,0\,1\,0 \\ 1\,1\,0\,0 \\ 1\,1\,1\,1 \end{pmatrix} \tag{7.1}$$

For fixed values of s and t, we may construct $OA(s^t, k, s, t)$ with $k > t + 1$ by Bush's construction [7] in Theorems 7.6 and 7.7.

THEOREM 7.6

If $q \geq 2$ is a prime power, then an $OA(q^t, q + 1, q, t)$ of index unity exists whenever $q \geq t - 1 \geq 0$.

PROOF   We first construct an $q^t \times q$ array whose columns are labeled with the elements α_j ($0 \leq j \leq q - 1$) of $\mathbb{F}_q$ and whose rows are labeled by the q^t polynomials with variable X over $\mathbb{F}_q$ of degree at most $t - 1$. Let those polynomials be denoted by $\phi_1, ..., \phi_{q^t}$. Then the entry in this array in the column labeled α_j and the row labeled by ϕ_i is defined to be $\phi_i(\alpha_j)$. Thus the q levels are denoted by the elements of $\mathbb{F}_q$.

We add one additional q-th column to this array, taking the level of this column in the row labeled ϕ_i to be the coefficient of X^{t-1} in ϕ_i.

Taking any t columns $0 \leq j_1 < j_2 < ... < j_t \leq q$ and a t-tuple of elements $u_1, u_2, ..., u_t$ in $\mathbb{F}_q$, we consider the following two cases.

Case 1 $j_t < q$. The system of linear equations with order t

$$a_{t-1}\alpha_{j_1}^{t-1} + a_{t-2}\alpha_{j_1}^{t-2} + ... + a_0 = u_1$$

$$a_{t-1}\alpha_{j_2}^{t-1} + a_{t-2}\alpha_{j_2}^{t-2} + ... + a_0 = u_2$$

$$...$$

$$a_{t-1}\alpha_{j_t}^{t-1} + a_{t-2}\alpha_{j_t}^{t-2} + ... + a_0 = u_t$$

has a unique solution $(a'_{t-1}, a'_{t-2}, ..., a'_0)$, which determines a unique polynomial $\phi_i = a'_{t-1}X^{t-1} + a'_{t-2}X^{t-2} + ... + a'_0$ such that $\phi_i(\alpha_{j_r}) = u_r$, $1 \leq r \leq t$.

Case 2 $j_t = q$. Similarly, the system of linear equations with order $t - 1$

$$a_{t-2}\alpha_{j_1}^{t-2} + ... + a_0 = u_1 - u_t\alpha_{j_1}^{t-1}$$

$$a_{t-2}\alpha_{j_2}^{t-2} + ... + a_0 = u_2 - u_t\alpha_{j_2}^{t-1}$$

$$\cdots$$

$$a_{t-2}\alpha_{j_{t-1}}^{t-2} + \ldots + a_0 = u_{t-1} - u_t\alpha_{j_{t-1}}^{t-1}$$

has a unique solution $(a'_{t-2}, \ldots, a'_0)$, which determines a unique polynomial $\phi_i = u_t X^{t-1} + a'_{t-2}X^{t-2} + \ldots + a'_0$ such that $\phi_i(\alpha_{j_r}) = u_r$, $1 \le r \le t-1$, and the coefficient of X^{t-1} in ϕ_i is u_t.   ▯

THEOREM 7.7

There exists an $OA(2^{3m}, 2^m + 2, 2^m, 3)$, where $m \ge 1$.

PROOF   We use the construction in Theorem 7.6 and the field $\mathbb{F}_{2^m}$ to obtain an $OA(2^{3m}, 2^m + 1, 2^m, 3)$, and adjoin one more column $\{(2^m + 1)-\text{th}\}$ for which the level in row ϕ_i is given by the coefficient of X in ϕ_i. {Note that $t = 3$ now, each ϕ_i has a degree of, at most, 2.}

Take any three columns $0 \le j_1 < j_2 < j_3 \le 2^m + 1$ and a 3-tuple of elements u_1, u_2, u_3 in $\mathbb{F}_{2^m}$. If $j_3 \le 2^m$, the verification of this case has already been given in Theorem 7.6. Hence, we need to consider only the case of $j_3 = 2^m + 1$.

Case 1 $j_2 < 2^m$, $j_3 = 2^m + 1$. The system of linear equations

$$a_2\alpha_{j_1}^2 + a_0 = u_1 - u_3\alpha_{j_1}$$

$$a_2\alpha_{j_2}^2 + a_0 = u_2 - u_3\alpha_{j_2}$$

has a unique solution (a'_2, a'_0) (note that $\alpha_{j_1}^2 \ne \alpha_{j_2}^2$) which determines a unique polynomial $\phi_i = a'_2 X^2 + u_3 X + a'_0$ such that $\phi_i(\alpha_{j_r}) = u_r$, $r = 1, 2$.

Case 2 $j_2 = 2^m$, $j_3 = 2^m + 1$. The equation

$$u_2\alpha_{j_1}^2 + u_3\alpha_{j_1} + a_0 = u_1$$

has a unique solution a'_0 which determines a unique polynomial $\phi_i = u_2 X^2 + u_3 X + a'_0$ such that $\phi_i(\alpha_{j_1}) = u_1$.   ▯

By using $OA(q^t, q+1, q, t)$ we can construct a t-fold perfect Cartesian A-code. Taking the set of source states as $\mathscr{S} = \mathbb{F}_q \bigcup \{\infty\}$ and the set of encoding rules as $\mathscr{E} = \{\phi_i | 1 \le i \le q^t\}$, which is the set of all polynomials of a degree that is, at most, $t-1$. The action of ϕ_i on $\mathscr{S}$ is defined by

$$\phi_i : \alpha_j \mapsto (\alpha_j, \phi_i(\alpha_j))$$

$$\infty \mapsto (\infty, u)$$

where u is the coefficient of X^{t-1} in ϕ_i.

Similarly, we can use $OA(2^{3m}, 2^m + 2, 2^m, 3)$ to construct a 3-fold perfect Cartesian A-code.

7.5 OA and Error-Correcting Codes

Let S be a set of symbols of size s, called the alphabet, and S^k be the set of all s^k vectors of length k. An error-correcting code, or simply a code, is any collection C of vectors in S^k. The vectors in C are called codewords.

We discuss the connections of orthogonal arrays with codes. These two subjects are very closely related, since we can regard the rows of an array as a code, or conversely we can use the codewords in a code as the rows of an array.

The Hamming weight $w(u)$ of a vector $u = (u_1, u_2, ..., u_k) \in S^k$ is defined to be the number of nonzero components of u. The Hamming distance $dist(u, v)$ between two vectors $u, v, \in S^k$ is defined to be the number of positions where they differ. We define the minimal distance d of a code C to be the minimal distance between distinct codewords:

$$d = \min_{u,v \in C, u \neq v} dist(u, v).$$

If C contains N codewords, then we say that it is a code of length k, size N, and minimal distance d over an alphabet of size s, denoted by $(k, N, d)_s$.

Let the alphabet S be the finite field $\mathbb{F}_q$, and we denote a code over $\mathbb{F}_q$ by (k, N, d). A code C of length k is said to be linear if the codewords are distinct and C is a vector subspace of $\mathbb{F}_q^{(k)}$. It implies that C has size $N = q^n$ for some nonnegative integer n, $0 \leq n \leq k$, called the dimension of the code. Similarly, an orthogonal array $OA(N, k, q, t)$ is said to be linear if its rows are distinct and form a vector subspace of $\mathbb{F}_q^{(k)}$. For simplicity, we mainly discuss linear orthogonal arrays and linear codes in this chapter, although some results also hold in the nonlinear case. The orthogonal arrays constructed in Theorems 7.6 and 7.7 are linear, since $\alpha_1 \phi_1 + \alpha_2 \phi_2$ is still a polynomial if ϕ_1 and ϕ_2 are polynomials and $\alpha_1, \alpha_2 \in \mathbb{F}_q$.

For a linear code the minimal distance d is equal to the minimal weight of all nonzero codewords:

$$d = \min_{u \in C, u \neq 0} w(u).$$

A linear code may be concisely specified by giving an $n \times k$ generator matrix G, whose rows form a basis of the code. The code then consists of all vectors $u = xG$, where x runs through $\mathbb{F}_q^{(n)}$. For example, the code corresponding to $OA(8, 4, 2, 3)$ in Equation (7.1) has the generator matrix

$$\begin{pmatrix} 1 & 0 & 0 & 1 \\ 0 & 1 & 0 & 1 \\ 0 & 0 & 1 & 1 \end{pmatrix}$$

An alternative way to specify a linear code is to give a parity check matrix H, which is a $(k - n) \times k$ matrix over $\mathbb{F}_q$ and its rows span the orthogonal space to the code. The code then consists of all vectors $u \in \mathbb{F}_q^{(k)}$ such that $Hu^T = 0$. A parity check matrix for the code in Equation (7.1) is

$$H = (1, 1, 1, 1).$$

Associated with any linear code C is another code called its dual, and denoted by $C^\perp$. It consists of all vectors $v \in \mathbb{F}_q^{(k)}$ such that

$$uv^T = 0 \ for \ all \ u \in C.$$

If C is a (k, q^n, d) linear code, then $C^\perp$ is a $(k, q^{n-k}, d^\perp)$ code for some number $d^\perp$. A generator matrix for C is a parity check matrix for $C^\perp$, and, vice versa, a parity check matrix for C is a generator matrix for $C^\perp$. We also have $(C^\perp)^\perp = C$.

We are interested in the relation between two parameters: the strength and the minimal distance.

PROPOSITION 7.11

Let A be an orthogonal array $OA(N, k, q, t)$ with entries from $\mathbb{F}_q$. Then any t columns of A are linearly independent over $\mathbb{F}_q$. Conversely, if A is an $N \times k$ matrix whose rows form a linear subspace of $\mathbb{F}_q^{(k)}$ and any t columns of A are linearly independent over $\mathbb{F}_q$, then A is an orthogonal array $OA(N, k, q, t)$.

PROOF   Take any t columns $v_1, v_2, ..., v_t$ of the orthogonal array $OA(N, k, q, t)$, and suppose

$$c_1 v_1 + c_2 v_2 + ... + c_t v_t = 0 \quad c_1, ..., c_t \in \mathbb{F}_q \qquad (7.2)$$

The matrix $(v_1 v_2 ... v_t)$ has a row $(1, 0, ..., 0)$; it implies that $c_1 = 0$ by Equation (7.2). Similarly, we have $c_2 = ... = c_t = 0$. This proves that $v_1, v_2, ..., v_t$ are linearly independent.

Suppose that A is an $N \times k$ matrix whose rows form a linear subspace of $\mathbb{F}_q^k$, and any t columns of A are linearly independent over $\mathbb{F}_q$. Then $N = q^n$ for some $0 < n \leq k$. Let G be an $n \times k$ generator for A, so that rows of A consist of all k-tuples ξG, where $\xi \in \mathbb{F}_q^{(n)}$. Choose t columns of A, and let G_1 be the corresponding $n \times t$ submatrix of G. Clearly, the columns of G_1 are linearly independent, otherwise the corresponding t columns of A would be linearly dependent. The number of times that a t-tuple z appears as a row in these t columns of A is equal to the number of ξ such that

$$\xi G_1 = z.$$

Since G_1 has rank t, this number is q^{n-t} for all z. Therefore, A is an orthogonal array of strength t.   □

PROPOSITION 7.12

If C is a (k, N, d) linear code over $\mathbb{F}_q$ with dual distance $d^\perp$, then the code-words of C form the rows of an $OA(N, k, q, d^\perp - 1)$ with entries in $\mathbb{F}_q$. Conversely, the rows of a linear $OA(N, k, q, t)$ over $\mathbb{F}_q$ form a (k, N, d) linear code over $\mathbb{F}_q$ with dual distance $d^\perp \geq t + 1$. If the orthogonal array has strength t but not $t + 1$, then $d^\perp$ is precisely $t + 1$.

PROOF   Suppose C is a code with the properties stated in the first part of the proposition, and let A be the array formed by the codewords. Any $d^\perp - 1$ columns of A must be linearly independent over $\mathbb{F}_q$, otherwise there would be a codeword of weight less that $d^\perp$ in the dual code, contradicting the hypothesis that $d^\perp$ is the minimal nonzero weight in the dual. By Proposition 7.11, A is an $OA(N, k, q, d^\perp - 1)$.

Conversely, Let C be the code associated with a linear $OA(N, k, q, t)$. By Proposition 7.11, any t columns of the array are linearly independent; and so there cannot be a codeword of weight t or less in $C^\perp$. If the array does not have strength $t + 1$, some $t + 1$ columns are dependent, and so there is a codeword of weight $t + 1$ in the dual code, hence $d^\perp = t + 1$. ꠸

Note that there is no direct relationship between d and t in the proposition, only between $d^\perp$ and t.

7.6   MDS Codes

PROPOSITION 7.13

Let C be a set of vectors over $\mathbb{F}_q$ of length k with minimal distance d and strength t. Then

$$q^t \leq |C| \leq q^{k-d+1} \tag{7.3}$$

where the right-hand side bound assumes that C has no repeated codeword.

PROOF   The left-hand inequality is obvious. By deleting any $d-1$ particular coordinates from all the vectors we obtain $|C|$ vectors of length $k - d + 1$, any two of which are different since the minimal distance of C is d. Therefore, the right-hand side inequality holds. The right-hand inequality of Equation (7.3) is called the Singleton bound. ꠸

A code (k, N, d) achieving the upper bound

$$|C| = q^{k-d+1}$$

is called a Maximal Distance Separable (**MDS**) code. For any linear (k, q^n, d) code the Singleton bound says that

$$d \leq k - n + 1 \qquad (7.4)$$

and a linear code is MDS code if and only if the equality in Equation (7.4) holds.

If the code C has the strength t and $|C| = q^t$, then the codewords of C form the rows of an $OA(q^t, k, q, t)$ of index unity. It is interesting that the set of MDS codes is nothing but the set of such orthogonal arrays of index unity.

THEOREM 7.8
Let C be a set of vectors over $\mathbb{F}_q$ of length k with minimal distance d and strength t. If the equality holds on either side of Equation (7.3), then it holds on both sides and C is simultaneously an $OA(q^t, k, q, t)$ of index unity and an MDS code with size q^t, minimal distance $d = k - t + 1$, dual distance $d^\perp = t + 1$, and $d + d^\perp = k + 2$.

PROOF   Suppose that $|C| = q^{k-d+1}$ first. The proof of Proposition 7.13 shows that if we delete any $d - 1$ coordinates, the shortened vectors are distinct, and so C has strength $t \geq k - d + 1$. Hence, $|C| = q^t$ by Equation (7.3). Conversely, suppose $|C| = q^t$. All codewords of C form the rows of a $q^t \times k$ array with strength t, hence any $q^t \times t$ subarray has q^t different rows, which implies that for any two codewords $u = (u_1, u_2, \cdots, u_k), v = (v_1, v_2, \cdots, v_k)$ of C, there is at least one different coordinate $u_i \neq v_i$ among any t-subset of coordinates. The only way for this to happen is $d(u, v) > k - t$. Thus $|C| = q^t$ if and only if $d \geq k - t + 1$, i.e., $t \geq k - d + 1$. It follows that $|C| = q^{k-d+1}$ by Equation (7.3). The strength of C cannot be $t + 1$, hence $d^\perp = t + 1$ by Proposition 7.12.   ⏹

Note that we have found a direct relation between d and t for MDS codes.

THEOREM 7.9
Let C be a linear $(k, q^t, k - t + 1)$ MDS code and (equivalently) a linear $OA(q^t, k, q, t)$. Then the dual $C^\perp$ is a linear $(k, q^{k-t}, t + 1)$ MDS code and linear $OA(q^{k-t}, k, q, k - t)$.

PROOF   This is an immediate consequence of Theorem 7.8.   ⏹

Combining Theorems 7.7 and 7.9 we have

THEOREM 7.10
There exists an $OA(2^{m[2^m - 1]}, 2^m + 2, 2^m, 2^m - 1)$ when $m \geq 1$.

The Reed–Solomon code is an important kind of MDS code. We introduce the concept of cyclic codes at first. A linear code C over $\mathbb{F}_q$ is said to be cyclic if whenever

$$(c_0, c_1, ..., c_{k-2}, c_{k-1}) \tag{7.5}$$

is a codeword, so is

$$(c_{k-1}, c_0, ..., c_{k-3}, c_{k-2}) \tag{7.6}$$

The same terminology can be applied to orthogonal arrays. An orthogonal array is cyclic if it is linear and if whenever Equation(7.5) is its row, so is Equation (7.6). The codeword in Equation (7.5) can be represented by the polynomial

$$c_0 + c_1 X + ... + c_{k-1} X^{k-1},$$

and the codeword in Equation (7.6) is represented by the polynomial

$$X(c_0 + c_1 X + ... + c_{k-1} X^{k-1}) \ mod(X^k - 1).$$

Thus the code C corresponds to an ideal $\mathcal{A}$ in the ring $\mathbb{F}_q[Z]/(X^k - 1)$, which is a principal ideal and has a generator polynomial $g(X)$ with the lowest degree in $\mathcal{A}$ and the leading coefficient 1. The set of all codewords of C are represented by the polynomial $a(X)g(X)$ where $a(X)$ runs through all polynomials with coefficients in $\mathbb{F}_q$ and degree not exceeding $k-1-deg\{g(X)\}$. It follows that the dimension of C is $k - deg\{g(X)\}$. The polynomial $g(X)$ is a factor of $X^k - 1$, otherwise the greatest common divisor of $g(X)$ and $X^k - 1$ belongs to $\mathcal{A}$ and has the degree less than $deg\{g(X)\}$ contradicting the definition of $g(X)$.

Let ξ be a primitive element of $\mathbb{F}_q$ and $k = q-1$. The cyclic code generated by the polynomial

$$g(X) = (X - \xi)(X - \xi^2)...(X - \xi^{d'-1}),$$

where $2 \leq d' \leq q-1$, has length $k = q-1$ and dimension $k - d' + 1$ as shown above. This code is called a Reed–Solomon code.

PROPOSITION 7.14
The Reed–Solomon code has minimal distance $d = d'$.

PROOF Let C denote the Reed–Solomon code defined above. Introduce the $(d' - 1) \times k$ matrix

$$H = \begin{pmatrix} 1 & \xi & \xi^2 & \cdots & \xi^{q-2} \\ 1 & \xi^2 & \xi^4 & \cdots & \xi^{2(q-2)} \\ & & \cdots & & \\ 1 & \xi^{d'-1} & \xi^{2(d'-1)} & \cdots & \xi^{(q-2)(d'-1)} \end{pmatrix}$$

which is the parity check matrix of C, since each code word of C

$$\sum_{i=0}^{q-2} c_i X^i = a(X)g(X) \quad mod(X^{q-1} - 1)$$

has $\xi^i (1 \leq i \leq d' - 1)$ as its zeros. Any $(d' - 1) \times (d' - 1)$ submatrix of H, as a Vandermonde matrix, is invertible, which implies $d \geq d'$. By the Singleton bound,

$$d \leq k - (k - d' + 1) + 1 = d',$$

this proves that $d = d'$. $\quad\square$

Put $n = k - d + 1 = q - d$; thus we have constructed cyclic Reed–Solomon codes with parameters $(q - 1, q^n, q - n)$ for all $n = 1, 2, ..., q - 1$. ($n = q - 1$ is a trivial case.)

It is known [24] that the length can be increased by 2, yielding codes with parameters

$$(q + 1, q^n, q - n + 2), \quad 1 \leq n \leq q + 1 \tag{7.7}$$

and that if $q = 2^m$ there are two cases when the length can be increased by 3, yielding extended Reed–Solomon code with parameters

$$(2^m + 2, 2^{3m}, 2^m) \tag{7.8}$$

and

$$(2^m + 2, 2^{m(2^m - 1)}, 4). \tag{7.9}$$

The dual distances for the codes in Equations (7.7), (7.8), (7.9) are respectively $n+1$, 4, and 2^m by Theorem 7.8, so the corresponding orthogonal arrays have parameters

$$OA(q^n, q + 1, q, n)$$

$$OA(2^{3m}, 2^m + 1, 2^m, 3)$$

$$OA(2^{m(2^m - 1)}, 2^m + 2, 2^m, 2^m - 1).$$

These are the same parameters as the array found in Reference [7] and described in Theorems 7.6, 7.7, and 7.10.

7.7   Comments

Orthogonal Latin squares of order 1 exist, but they are not interesting. It is not difficult to see that there do not exist orthogonal Latin squares of order 2. Over 200 years ago, the mathematician Euler conjectured that there do not exist orthogonal Latin squares of order n if $n \equiv 2 \ (mod \ 4)$. Euler's conjecture was proved true for order 6 by Tarry in 1900, essentially by means of an exhaustive search. On the other hand, for all integers $n \neq 2, 6$, there do exist orthogonal Latin squares of order n. This disproof of Euler's conjecture was published in the late 1950s by Bose and Shrikhande [4] and Bose et al. [5], and it was reported on the front page of *The New York Times*. The proof of this result in this chapter is not the original one, but is simplified by using Wilson's construction [57] (see Stinson [49]).

For more constructions of orthogonal arrays the readers may refer to the book, *Orthogonal Arrays, Theory and Application* by A.S. Hedayat, N.J.A. Sloane, and John Stufken [14].

Some material (the text pp 131-134, pp. 140-144, and pp. 147-150) of this chapter is cited from the text pp. 38-39 and pp. 96-98 of [14], and pp. 133-135, p. 143, and pp. 151-152 of [49] with kind permission of Springer Science and Business Media.

7.8   Exercises

7.1 Prove that the minimal distance of a linear code is equal to the minimal weight of all nonzero codewords.

7.2 Let P be a $(k-n) \times k$ parity check matrix for a linear code C over $\mathbb{F}_q$. Show that C has minimal distance d if and only if every $d-1$ column of P are linearly independent over $\mathbb{F}_q$ and some d columns are linearly dependent.

7.3 Let C be a (k, q^n, d) linear code over $\mathbb{F}_q$. Prove: (a) the dual code $C^\perp$ has length k, dimension $k-n$, and minimal distance $d^\perp$ for some number $d^\perp$, i.e., $C^\perp$ is a $(k, q^{k-n}, d^\perp)$ code; (b) a generator matrix for C is a parity check matrix for $C^\perp$, a parity check matrix for C is a generator matrix for $C^\perp$; and (c) $(C^\perp)^\perp = C$.

7.4 Let C be the trivial (k, q, k) repetition code with a generator matrix consisting of the single vector $(1, 1, ..., 1)$ over $\mathbb{F}_q$, which forms a $OA(q, k, q, 1)$. Show that the dual code $C^\perp$ is a zero-sum code with parameters $(k, q^{k-1}, 2)$ whose codewords form a zero-sum orthogonal array $OA(q^{k-1}, k, q, k-1)$.

7.5 Construct a $MOLS(8, 7)$ using Theorem 7.6.

Chapter 8

A-Codes from Finite Geometries

The geometry of classical groups (such as symplectic groups, unitary groups, and orthogonal groups) over finite fields can be used to construct authentication codes. As examples, two schemes from the geometry of symplectic groups and one scheme from the geometry of unitary groups are presented in this chapter. For the convenience of readers, the fundamental knowledge about symplectic spaces, which is necessary for use in this chapter, is given in Section 8.1.

8.1 Symplectic Spaces over Finite Fields

Let $\mathbb{F}_q$ be a field with q elements, where q may be odd or even. Define the matrix

$$K = \begin{pmatrix} 0 & I^{(v)} \\ -I^{(v)} & 0 \end{pmatrix} \tag{8.1}$$

of order $n = 2v$, where $I^{(v)}$ is the identity matrix of order v. The symplectic group of degree $2v$ over $\mathbb{F}_q$, denoted by $Sp_{2v}(\mathbb{F}_q)$, is defined to be the set of matrices

$$Sp_{2v}(\mathbb{F}_q) = \{T \in GL_{2v}(\mathbb{F}_q) \mid TKT^t = K\}$$

with matrix multiplication as its group operation, where $GL_{2v}(\mathbb{F}_q)$ is the group consisting of all $2v \times 2v$ nonsingular matrices over $\mathbb{F}_q$ and T^t is the transpose of T.

Each matrix T of $Sp_{2v}(\mathbb{F}_q)$ defines a transformation on the vector space $\mathbb{F}_q^{(2v)}$ by

$$\begin{aligned} \mathbb{F}_q^{(2v)} &\longrightarrow \mathbb{F}_q^{(2v)} \\ (x_1, x_2, \cdots, x_{2v}) &\longmapsto (x_1, x_2, \cdots, x_{2v})T, \end{aligned}$$

which is called a symplectic transformation. The vector space $\mathbb{F}_q^{(2v)}$, together with the above group action of the symplectic group $Sp_{2v}(\mathbb{F}_q)$, is called the $2v$-dimensional symplectic space over $\mathbb{F}_q$.

Let P be an m-dimensional subspace of $\mathbb{F}_q^{(2v)}$. We use the same letter P to denote an $m \times 2v$ matrix of rank m whose rows form a basis of P. It is easy to see that AP and P denote the same subspace where A is an $m \times m$ nonsingular matrix over $\mathbb{F}_q$. We are interested in the transitive relation between subspaces of $\mathbb{F}_q^{(2v)}$ under symplectic transformations.

DEFINITION 8.1   *An $n \times n$ matrix $A = (a_{ij})_{1 \leqslant i,j \leqslant n}$ is said to be alternate if $a_{ij} = -a_{ji}$ for $i \neq j$, $1 \leqslant i,j \leqslant n$ and $a_{ii} = 0$ for $i = 1, 2, \cdots, n$.*

DEFINITION 8.2   *Let S_1 and S_2 be two $n \times n$ matrices over $\mathbb{F}_q$. If there is an $n \times n$ nonsingular matrix Q over $\mathbb{F}_q$ such that $QS_1Q^t = S_2$, then S_1 is said to be cogredient to S_2.*

THEOREM 8.1

Let A be an $n \times n$ alternate matrix over $\mathbb{F}_q$. The rank of A is necessarily even. Furthermore, if A is of rank $2s$ ($\leqslant n$), then A is cogredient to

$$\begin{pmatrix} 0 & I^{(s)} & \\ -I^{(s)} & 0 & \\ & & 0^{(n-2s)} \end{pmatrix},$$

and the integer s is called the index of A. (The blank space represents "zeroes". Similarly, in the following.)

PROOF   Use induction on n. For $n = 1$, $A = (0)$, the theorem holds. For $n = 2$, A has the form

$$A = \begin{pmatrix} 0 & a_{12} \\ -a_{12} & 0 \end{pmatrix}.$$

If $a_{12} = 0$, then the theorem also holds; if $a_{12} \neq 0$, then A is cogredient to

$$\begin{pmatrix} 1 & 0 \\ 0 & a_{12}^{-1} \end{pmatrix} A \begin{pmatrix} 1 & 0 \\ 0 & a_{12}^{-1} \end{pmatrix}^t = \begin{pmatrix} 0 & 1 \\ 1 & 0 \end{pmatrix}.$$

Now suppose that $n > 2$ and that the theorem holds for $m < n$. If $A = 0$, then the theorem is true. Suppose that $A \neq 0$, then there is a $a_{ij} \neq 0$, $i < j$. Interchanging the first row and the i-th row of A and the first column and the i-th column of A simultaneously, we obtain an alternate matrix A_1 which is cogredient to A and whose element at the $(1,j)$ position is nonzero. Then interchanging the second row and the j-th row of A_1 and the second column and the j-th column of A_1 simultaneously we obtain an alternate matrix A_2 which is cogredient to A_1 and also to A and whose element at the $(1,2)$ position is nonzero. Thus we can assume that $a_{12} \neq 0$ in A. Furthermore, A is cogredient to

$$\begin{pmatrix} 1 & \\ & a_{12}^{-1} \\ & & I^{(n-2)} \end{pmatrix} A \begin{pmatrix} 1 & \\ & a_{12}^{-1} \\ & & I^{(n-2)} \end{pmatrix}^t,$$

and we can assume that $a_{12} = 1$. Then A is cogredient to

$$\begin{pmatrix} 1 & & & & & \\ 0 & 1 & & & & \\ a_{32} & a_{13} & 1 & & & \\ \vdots & \vdots & 0 & \ddots & & \\ \vdots & \vdots & \vdots & \ddots & \ddots & \\ a_{n2} & a_{1n} & 0 & \cdots & 0 & 1 \end{pmatrix} A \begin{pmatrix} 1 & & & & & \\ 0 & 1 & & & & \\ a_{32} & a_{13} & 1 & & & \\ \vdots & \vdots & 0 & \ddots & & \\ \vdots & \vdots & \vdots & \ddots & \ddots & \\ a_{n2} & a_{1n} & 0 & \cdots & 0 & 1 \end{pmatrix}^t = \begin{pmatrix} 0 & 1 & \\ -1 & 0 & \\ & & A_0^{(n-2)} \end{pmatrix}$$

where A_0 is an $(n-2) \times (n-2)$ alternate matrix, which is cogredient to

$$\begin{pmatrix} 0 & I^{(s-1)} & \\ -I^{(s-1)} & 0 & \\ & & 0^{(n-2s)} \end{pmatrix}.$$

Since we have

$$\begin{pmatrix} 1 & & & & & \\ & 0 & 1 & & & \\ & & 0 & 1 & & \\ & & & \ddots & & \\ & & & & 0 & 1 \\ & 1 & 0 & \cdots & & 0 \\ & & & & & I^{(s-1)} \\ & & & & & & 0^{(n-2s)} \end{pmatrix} \begin{pmatrix} 0 & 1 & & & \\ -1 & 0 & & & \\ & & 0 & I^{(s-1)} & \\ & & -I^{(s-1)} & 0 & \\ & & & & 0^{(n-2s)} \end{pmatrix}$$

$$\times \begin{pmatrix} 1 & & & & & \\ & 0 & 1 & & & \\ & & 0 & 1 & & \\ & & & \ddots & & \\ & & & & 0 & 1 \\ & 1 & 0 & \cdots & & 0 \\ & & & & & I^{(s-1)} \\ & & & & & & 0^{(n-2s)} \end{pmatrix}^t = \begin{pmatrix} 0 & I^{(s)} & \\ -I^{(s)} & 0 & \\ & & 0^{(n-2s)} \end{pmatrix},$$

the theorem is proved. $\qquad\qquad\qquad\qquad\qquad\qquad\qquad\qquad\qquad\quad$ ▯

The matrix K defined in Equation (8.1) is a nonsingular alternative with index $v = n/2$. Suppose that P is an m-dimensional subspace of $\mathbb{F}_q^{(2v)}$, then PKP^t is an $m \times m$ alternative matrix. By Theorem 8.1, let the rank of PKP^t be $2s$, then we call the vector subspace P a subspace of type (m, s).

Clearly, $s \leqslant v$ and $2s \leqslant m$. In particular, subspaces of type $(m, 0)$ are called m-dimensional totally isotropic subspaces, and subspaces of type $(2s, s)$ are called $2s$-dimensional nonisotropic subspaces. It is clear that a subspace P is totally isotropic if and only if $PKP^t = 0$, and it is nonisotropic if and only if PKP^t is nonsingular.

Clearly,

$$(x_1, \cdots, x_v, x_{v+1}, \cdots, x_{2v})K(x_1, \cdots, x_v, x_{v+1}, \cdots, x_{2v})^t$$
$$= (-x_{v+1}, \cdots, -x_{2v}, x_1, \cdots, x_v)(x_1, \cdots, x_v, x_{v+1}, \cdots, x_{2v})^t$$
$$= 0,$$

hence, all 1-dimensional subspaces are totally isotropic. If $s \leqslant v$, then

$$
\begin{array}{cccc}
s & v-s & s & v-s
\end{array}
$$
$$
\left(\begin{array}{cccc} I^{(s)} & 0 & 0 & 0 \end{array} \right)
\tag{8.2}
$$

is totally isotropic of dimension s, and

$$
\begin{array}{cccc}
s & v-s & s & v-s
\end{array}
$$
$$
\left(\begin{array}{cccc} I^{(s)} & 0 & 0 & 0 \\ 0 & 0 & I^{(s)} & 0 \end{array} \right)
\tag{8.3}
$$

is a nonisotropic subspace of dimension $2s$. If $2s \leqslant m \leqslant v + s$, then

$$
\begin{array}{cccccc}
s & m-2s & v+s-m & s & m-2s & v+s-m
\end{array}
$$
$$
\left(\begin{array}{cccccc} I^{(s)} & 0 & 0 & 0 & 0 & 0 \\ 0 & 0 & 0 & I^{(s)} & 0 & 0 \\ 0 & I^{(m-2s)} & 0 & 0 & 0 & 0 \end{array} \right)
\tag{8.4}
$$

is a subspace of type (m, s).

Two vectors x and y of $\mathbb{F}_q^{(2v)}$ are said to be orthogonal (with respect to K) if $xKy^t = 0$. Clearly, if $xKy^t = 0$, then $yKx^t = 0$. Every vector is orthogonal to itself, i.e., $xKx^t = 0$ for every vector x of $\mathbb{F}_q^{(2v)}$.

Let P be an m-dimensional subspace of $\mathbb{F}_q^{(2v)}$. Denote by $P^\perp$ the set of vectors which are orthogonal to every vector of P, i.e.,

$$P^\perp = \{y \in \mathbb{F}_q^{(2v)} \mid yKx^t = 0 \quad \text{for all} \quad x \in P\}.$$

Obviously, $P^\perp$ is a $(2v - m)$-dimensional subspace of $\mathbb{F}_q^{(2v)}$ and is called the dual subspace of P. For example, the dual subspace of Equation (8.2) is

$$
\begin{array}{cccc}
s & v-s & s & v-s
\end{array}
$$
$$
\left(\begin{array}{cccc} I^{(s)} & 0 & 0 & 0 \\ 0 & I^{(s)} & 0 & 0 \\ 0 & 0 & 0 & I^{(s)} \end{array} \right),
$$

and the dual subspace of Equation (8.3) is

$$
\begin{array}{cccc}
s & v-s & s & v-s
\end{array}
$$
$$
\begin{pmatrix}
0 & I^{(v-s)} & 0 & 0 \\
0 & 0 & 0 & I^{(v-s)}
\end{pmatrix}.
$$

The dual subspace of Equation (8.4), which is of type (m, s), is

$$
\begin{array}{cccccc}
s & m-2s & v+s-m & s & m-2s & v+s-m
\end{array}
$$
$$
\begin{pmatrix}
0 & I^{(m-2s)} & 0 & 0 & 0 & 0 \\
0 & 0 & I^{(v+s-m)} & 0 & 0 & 0 \\
0 & I^{(m-2s)} & 0 & 0 & 0 & I^{(v+s-m)}
\end{pmatrix},
$$

and is of type $(2v - m, v + s - m)$.

THEOREM 8.2

A subspace P is totally isotropic if and only if $P \subset P^\perp$; and P is nonisotropic if and only if $P \cap P^\perp = \{0\}$.

PROOF By definition a subspace P is totally isotropic if $PKP^t = 0$, and $PKP^t = 0$ if and only if $P \subset P^\perp$, whence the first assertion is proved.

Notice that

$$
\begin{pmatrix} P \\ P^\perp \end{pmatrix} K \begin{pmatrix} P \\ P^\perp \end{pmatrix}^t = \begin{pmatrix} PKP^t & 0 \\ 0 & P^\perp K (P^\perp)^t \end{pmatrix}.
$$

If $P \cap P^\perp = \{0\}$, then

$$
\begin{pmatrix} P \\ P^\perp \end{pmatrix}
$$

is a $2v \times 2v$ nonsingular matrix, and hence PKP^t is nonsingular, i.e., P is nonisotropic. Conversely, suppose that there is a non-zero vector $x \in P \cap P^\perp$. Let P be an m-dimensional subspace, then there is a $1 \times m$ vector $\xi \neq 0$ such that $x = \xi P$. Thus, $\xi P K P^t = x K P^t = 0$; it follows that P is not nonisotropic. □

The following lemma is crucial for studying the transitivity properties of $Sp_{2v}(\mathbb{F}_q)$ on the set of subspaces of $\mathbb{F}_q^{(2v)}$.

LEMMA 8.1
Let P be a subspace of type (m, s) in $\mathbb{F}_q^{(2v)}$. Then $2s \leqslant m \leqslant v + s$ and there is an $m \times m$ nonsingular matrix Q and a $(2v - m) \times 2v$ matrix Z such that

$$\begin{pmatrix} QP \\ Z \end{pmatrix} K \begin{pmatrix} QP \\ Z \end{pmatrix}^t = \begin{pmatrix} \begin{matrix} 0 & I^{(s)} \\ -I^{(s)} & 0 \end{matrix} & & \\ & \begin{matrix} 0 & I^{(m-2s)} \\ -I^{(m-2s)} & 0 \end{matrix} & \\ & & \begin{matrix} 0 & I^{(\sigma)} \\ -I^{(\sigma)} & 0 \end{matrix} \end{pmatrix} \tag{8.5}$$

where $\sigma = v + s - m$.

PROOF   It follows trivially from the definition that $2s \leq m$. Since P is a subspace of type (m, s), PKP^t is an $m \times m$ alternate matrix of rank $2s$. By Theorem 8.1, there is an $m \times m$ nonsingular matrix Q such that

$$Q(PKP^t)Q^t = \begin{pmatrix} \begin{matrix} 0 & I^{(s)} \\ -I^{(s)} & 0 \end{matrix} & \\ & I^{(m-2s)} \end{pmatrix}.$$

There is a $(2v - m) \times 2v$ matrix Z_1 such that

$$\begin{pmatrix} QP \\ Z_1 \end{pmatrix}$$

is nonsingular. We may write

$$\begin{pmatrix} QP \\ Z_1 \end{pmatrix} K \begin{pmatrix} QP \\ Z_1 \end{pmatrix}^t = \begin{pmatrix} 0 & I^{(s)} & 0 & K_{14} \\ -I^{(s)} & 0 & 0 & K_{24} \\ 0 & 0 & 0^{(m-2s)} & K_{34} \\ -K_{14}^t & -K_{24}^t & -K_{34}^t & K_{44} \end{pmatrix},$$

where K_{44} is a $(2v - m) \times (2v - m)$ alternate matrix. Then we have

$$\begin{pmatrix} I^{(s)} & & & \\ & I^{(s)} & & \\ & & I^{(s)} & \\ K_{24}^t & -K_{14}^t & 0 & I^{(2v-m)} \end{pmatrix} \begin{pmatrix} 0 & I^{(s)} & 0 & K_{14} \\ -I^{(s)} & 0 & 0 & K_{24} \\ 0 & 0 & 0^{(m-2s)} & K_{34} \\ -K_{14}^t & -K_{24}^t & -K_{34}^t & K_{44} \end{pmatrix}$$
$$\times \begin{pmatrix} I^{(s)} & & & \\ & I^{(s)} & & \\ & & I^{(s)} & \\ K_{24}^t & -K_{14}^t & 0 & I^{(2v-m)} \end{pmatrix}^t = \begin{pmatrix} 0 & I^{(s)} & 0 & 0 \\ -I^{(s)} & 0 & 0 & 0 \\ 0 & 0 & 0 & K_{34} \\ 0 & 0 & -K_{34}^t & L_{44} \end{pmatrix},$$

where $L_{44} = K_{44} + K_{24}^t K_{14} - K_{14}^t K_{24}$. Since K is nonsingular, K_{34} is of rank $m - 2s$. Thus, $m - 2s \leqslant 2v - m$ which gives $m \leqslant v + s$; and there is a $(2v - m) \times (2v - m)$ nonsingular matrix B such that

$$K_{34}B^t = \left(I^{(m-2s)}\ 0^{(2\sigma)}\right).$$

Then

$$
\begin{pmatrix} I & & & \\ & I & & \\ & & I & \\ & & & B \end{pmatrix}
\begin{pmatrix} 0 & I^{(s)} & & \\ -I^{(s)} & 0 & & \\ & & 0 & K_{34} \\ & & -K_{34}^t & L_{44} \end{pmatrix}
\begin{pmatrix} I & & & \\ & I & & \\ & & I & \\ & & & B \end{pmatrix}^t
$$

$$
= \begin{pmatrix} 0 & I^{(s)} & & & \\ -I^{(s)} & 0 & & & \\ & & 0 & I^{(m-2s)} & 0 \\ & & -I^{(m-2s)} & K_{44}' & K_{45} \\ & & 0 & -K_{45}^t & K_{55} \end{pmatrix},
$$

where K_{44}' is an $(m-2s) \times (m-2s)$ alternate matrix and K_{55} is a $2\sigma \times 2\sigma$ alternate matrix. Let T be the matrix obtained from K_{44}' by replacing all elements below the main diagonal by zero. Then

$$K_{44}' = T - T^t$$

and

$$
\begin{pmatrix} I^{(s)} & & & & \\ & I^{(s)} & & & \\ & & I^{(m-2s)} & & \\ & & -T & I^{(m-2s)} & \\ & & K_{45}^t & & I^{(2\sigma)} \end{pmatrix}
\begin{pmatrix} 0 & I^{(s)} & & & \\ -I^{(s)} & 0 & & & \\ & & 0 & I^{(m-2s)} & \\ & & -I^{(m-2s)} & K_{44}' & K_{45} \\ & & & -K_{45}' & K_{55} \end{pmatrix}
$$

$$
\times \begin{pmatrix} I^{(s)} & & & & \\ & I^{(s)} & & & \\ & & I^{(m-2s)} & & \\ & & -T & I^{(m-2s)} & \\ & & K_{45}^t & & I^{(2\sigma)} \end{pmatrix}^t
= \begin{pmatrix} 0 & I^{(s)} & & & \\ -I^{(s)} & 0 & & & \\ & & -I^{(m-2s)} & 0 & 0 \\ & & & 0 & K_{55} \end{pmatrix}.
$$

Since K is nonsingular, K_{55} is nonsingular. By Theorem 8.1 there is a $2\sigma \times 2\sigma$ nonsingular matrix R such that

$$RK_{55}R^t = \begin{pmatrix} 0 & I^{(\sigma)} \\ -I^{(\sigma)} & 0 \end{pmatrix}.$$

Write

$$
\begin{pmatrix} I^{(s)} & & & & \\ & I^{(s)} & & & \\ & & I^{(m-2s)} & & \\ & & & I^{(m-2s)} & \\ & & & & R \end{pmatrix}
\begin{pmatrix} I^{(s)} & & & & \\ & I^{(s)} & & & \\ & & I^{(m-2s)} & & \\ & & -T & I^{(m-2s)} & \\ & & K_{45}^t & & I^{(2\sigma)} \end{pmatrix}
$$

$$
\times \begin{pmatrix} I^{(s)} & & & \\ & I^{(s)} & & \\ & & I^{(m-2s)} & \\ & & & B \end{pmatrix}
\begin{pmatrix} I^{(s)} & & & \\ & I^{(s)} & & \\ & & I^{(m-2s)} & \\ & K_{24}^t & -K_{14}^t & I^{(2v-m)} \end{pmatrix}
= \begin{pmatrix} I^{(m)} \\ L_1 & L_2 \end{pmatrix},
$$

and let

$$\begin{pmatrix} I^{(m)} & \\ L_1 & L_2 \end{pmatrix} \begin{pmatrix} QP \\ Z_1 \end{pmatrix} = \begin{pmatrix} QP \\ Z \end{pmatrix},$$

then the equality in Equation (8.5) is proved. ▯

THEOREM 8.3
Subspaces of type (m, s) exist in the $2v$-dimensional symplectic space if and only if $2s \leqslant m \leqslant v + s$.

PROOF   Lemma 8.1 asserts that if there is a subspace of type (m, s) in the $2v$-dimensional symplectic space, then $2s \leqslant m \leqslant v + s$. Conversely, if $2s \leqslant m \leqslant v + s$, then Equation (8.4) is a subspace of type (m, s). ▯

The following transitivity theorem of subspaces of the same type under $Sp_{2v}(\mathbb{F}_q)$ also follows from Lemma 8.1.

THEOREM 8.4
Let P_1 and P_2 be two m-dimensional subspaces of $\mathbb{F}_q^{(2v)}$. Then there is a $T \in Sp_{2v}(\mathbb{F}_q)$ such that $P_1 = AP_2T$, where A is an $m \times m$ nonsingular matrix, if and only if P_1 and P_2 are of the same type. In other words, $Sp_{2v}(\mathbb{F}_q)$ acts transitively on each set of subspaces of the same type.

PROOF   Suppose that there is a $T \in Sp_{2v}(\mathbb{F}_q)$ such that $P_1 = AP_2T$, where A is an $m \times m$ nonsingular matrix. Then

$$P_1 K P_1^t = AP_2 T K T^t P_2^t A^t = AP_2 K P_2^t A^t,$$

i.e., $P_1 K P_1^t$ and $P_2 K P_2^t$ are cogredient. Hence P_1 and P_2 are of the same type.

Conversely, suppose P_1 and P_2 are of the same type (m, s). By Lemma 8.1 there are $m \times m$ nonsingular matrices Q_1 and Q_2, and $(2v - m) \times 2v$ matrices Z_1 and Z_2 such that both

$$\begin{pmatrix} Q_1 P_1 \\ Z_1 \end{pmatrix} K \begin{pmatrix} Q_1 P_1 \\ Z_1 \end{pmatrix}^t, \qquad \begin{pmatrix} Q_2 P_2 \\ Z_2 \end{pmatrix} K \begin{pmatrix} Q_2 P_2 \\ Z_2 \end{pmatrix}^t$$

are equal to the right-hand side of Equation (8.5). Thus

$$T = \begin{pmatrix} Q_2 P_2 \\ Z_2 \end{pmatrix}^{-1} \begin{pmatrix} Q_1 P_1 \\ Z_1 \end{pmatrix} \in Sp_{2v}(\mathbb{F}_q).$$

Therefore,

$$\begin{pmatrix} Q_2 P_2 \\ Z_2 \end{pmatrix} T = \begin{pmatrix} Q_1 P_1 \\ Z_1 \end{pmatrix},$$

and

$$P_1 = Q_1^{-1} Q_2 P_2 T.$$

☐

Furthermore, we have the following lemma.

LEMMA 8.2

Let P_1 and P_2 be two $m \times 2v$ matrices of rank m. Then there exists an element $T \in Sp_{2v}(\mathbb{F}_q)$ such that $P_1 = P_2 T$ if and only if $P_1 K P_1^t = P_2 K P_2^t$.

PROOF   Suppose that there is a $T \in Sp_{2v}(\mathbb{F}_q)$ such that $P_1 = P_2 T$, then

$$P_1 K P_1^t = P_2 T K T^t P_2^t = P_2 K P_2^t.$$

Now suppose that $P_1 K P_1^t = P_2 K P_2^t$. Then in the proof of Lemma 8.1 we can take the same $m \times m$ nonsingular matrix Q and two $(2v - m) \times 2v$ matrices Z_1 and Z_2 such that both

$$\begin{pmatrix} QP_1 \\ Z_1 \end{pmatrix} K \begin{pmatrix} QP_1 \\ Z_1 \end{pmatrix}^t \quad \text{and} \quad \begin{pmatrix} QP_2 \\ Z_2 \end{pmatrix} K \begin{pmatrix} QP_2 \\ Z_2 \end{pmatrix}^t$$

are equal to the right-hand side of Equation (8.5). Then

$$\begin{pmatrix} QP_1 \\ Z_1 \end{pmatrix} K \begin{pmatrix} QP_1 \\ Z_1 \end{pmatrix}^t = \begin{pmatrix} QP_2 \\ Z_2 \end{pmatrix} K \begin{pmatrix} QP_2 \\ Z_2 \end{pmatrix}^t.$$

Thus,

$$T = \begin{pmatrix} QP_2 \\ Z_2 \end{pmatrix}^{-1} \begin{pmatrix} QP_1 \\ Z_1 \end{pmatrix} \in Sp_{2v}(\mathbb{F}_q)$$

and

$$P_1 = P_2 T.$$

☐

Denote by $\mathcal{M}(m, s; 2v)$ the set of subspaces of $\mathbb{F}_q^{(2v)}$ with type (m, s) and by $N(m, s; 2v)$ the number of subspaces of $\mathbb{F}_q^{(2v)}$ with type (m, s), i.e.,

$$N(m, s; 2v) = |\mathcal{M}(m, s; 2v)|.$$

The remaining part of this section is devoted to the calculation of $N(m, s; 2v)$. In the following we always assume that

$$2s \leqslant m \leqslant v + s.$$

Let

$$M(m,s) = \begin{pmatrix} 0 & I^{(s)} & \\ -I^{(s)} & 0 & \\ & & 0^{(m-2s)} \end{pmatrix}. \tag{8.6}$$

If P is a subspace of type (m,s), then by Theorem 8.1 we can assume that

$$PKP^t = M(m,s). \tag{8.7}$$

Denote by $n(m,s;2v)$ the number of $m \times 2v$ matrices P of rank m which satisfy Equation (8.7). Then we have

LEMMA 8.3
Let $2s \leqslant m \leqslant v + s$. Then

$$n(m,s;2v) = |Sp_{2v}(\mathbb{F}_q)| \cdot |GL_{m-2s}(\mathbb{F}_q)| q^{2s(m-2s)} N(m,s;2v).$$

PROOF   Let P_1 and P_2 be two $m \times 2v$ matrices of rank m which satisfy Equation (8.7). If P_1 and P_2 represent the same subspace, then there is an $m \times m$ nonsingular matrix Q such that

$$P_1 = QP_2.$$

From $P_1 K P_1^t = P_2 K P_2^t = M(m,s)$, we deduce

$$QM(m,s)Q^t = M(m,s). \tag{8.8}$$

The set of all $m \times m$ nonsingular matrices Q which satisfy Equation (8.8) forms a group, denoted by $\mathcal{G}(m,s)$. For $Q \in \mathcal{G}(m,s)$, $Q \neq I$ implies $QP_2 \neq P_2$. Hence,

$$n(m,s;2v) = |\mathcal{G}(m,s)|N(m,s;2v). \tag{8.9}$$

Let

$$K_1 = \begin{pmatrix} 0 & I^{(s)} \\ -I^{(s)} & 0 \end{pmatrix}$$

and write Q in block forms

$$Q = \begin{pmatrix} A^{(2s)} & B \\ C & D^{(m-2s)} \end{pmatrix}.$$

From Equation (8.8) we deduce

$$AK_1 A^t = K_1, \quad AK_1 C^t = 0,$$

and

$$CK_1 C^t = 0.$$

It follows that $A \in Sp_{2v}(\mathbb{F}_q)$ and $C = 0$. That is, $\mathcal{G}(m, s)$ consists of all $m \times m$ matrices of the form

$$\begin{pmatrix} A & B \\ 0 & D \end{pmatrix},$$

where $A \in Sp_{2v}(\mathbb{F}_q)$, B is any $2s \times (m - 2s)$ matrix, and $D \in GL_{m-2s}(\mathbb{F}_q)$. Consequently,

$$|\mathcal{G}(m, s)| = |Sp_{2v}(\mathbb{F}_q)| \cdot |GL_{m-2s}(\mathbb{F}_q)| q^{2s(m-2s)} \tag{8.10}$$

and Lemma 8.2 follows from Equations (8.9) and (8.10). ⬜

Clearly,

$$|Sp_{2v}(\mathbb{F}_q)| = n(2s, s; 2s)$$

and

$$|GL_{m-2s}(\mathbb{F}_q)| = (q^{m-2s} - 1)(q^{m-2s} - q) \cdots (q^{m-2s} - q^{m-2s-1})$$
$$= q^{(m-2s-1)(m-2s)/2} \prod_{i=1}^{m-2s} (q^i - 1). \tag{8.11}$$

Thus, to compute $N(m, s; 2v)$, it is sufficient to compute $n(m, s; 2v)$.

Let us introduce the following notations. Let A be an $m \times n$ matrix and $0 \leqslant r \leqslant \min(m, n)$. Denote by A_r the upper-left corner $r \times r$ submatrix of A. Now let $2s \leqslant m \leqslant v + s$ and $0 \leqslant r_1, r_2, r_1 + r_2 \leqslant m$. Let $n(r_1)$ be the number of $r_1 \times 2v$ matrices P_1 of rank r_1 such that $P_1 K P_1^t = M(m, s)_{r_1}$. For such a fixed P_1 let $n(P_1, r_2)$ be the number of $r_2 \times 2v$ matrices P_2 such that

$$P = \begin{pmatrix} P_1 \\ P_2 \end{pmatrix}$$

is of rank $r_1 + r_2$ and $PKP^t = M(m, s)_{r_1+r_2}$. Then we have

LEMMA 8.4

Let $2s \leqslant m \leqslant v + s$ and $0 \leqslant r_1, r_2, r_1 + r_2 \leqslant m$. Then $n(P_1, r_2)$ is independent of the choice of P_1, and

$$n(r_1 + r_2) = n(r_1)n(P_1, r_2)$$

for any choice of P_1.

PROOF Let Q_1 be another $r_1 \times 2v$ matrix of rank r_1 such that $Q_1 K Q_1^t = M(m, s)_{r_1}$. Then $P_1 K P_1^t = Q_1 K Q_1^t$. By Lemma 8.2 there is an element $T \in Sp_{2v}(\mathbb{F}_q)$ such that $Q_1 = P_1 T$. Thus, if P_2 is an $r_2 \times 2v$ matrix such that

$$P = \begin{pmatrix} P_1 \\ P_2 \end{pmatrix}$$

is of rank $r_1 + r_2$ and $PKP^t = M(m,s)_{r_1+r_2}$, then P_2T is also an $r_2 \times 2v$ matrix such that

$$Q = \begin{pmatrix} Q_1 \\ P_2T \end{pmatrix} = \begin{pmatrix} P_1 \\ P_2 \end{pmatrix} T$$

is of rank $r_1 + r_2$ and $QKQ^t = M(m,s)_{r_1+r_2}$. Conversely, if Q_2 is an $r_2 \times 2v$ matrix such that

$$Q = \begin{pmatrix} Q_1 \\ Q_2 \end{pmatrix}$$

is of rank $r_1 + r_2$ and $QKQ^t = M(m,s)_{r_1+r_2}$, then Q_2T^{-1} is an $r_2 \times 2v$ matrix such that

$$P = \begin{pmatrix} P_1 \\ Q_2T^{-1} \end{pmatrix} = \begin{pmatrix} Q_1 \\ Q_2 \end{pmatrix} T^{-1}$$

is of rank $r_1 + r_2$ and $PKP^t = M(m,s)_{r_1+r_2}$. Hence, the lemma follows. $\quad\Box$

Now we are ready to compute $n(m,s;2v)$. Let us start with a special case.

LEMMA 8.5
Let $s \leqslant v$, then

$$n(s,0;2v) = q^{\frac{s(s-1)}{2}} \prod_{i=v-s+1}^{v} (q^{2i} - 1).$$

PROOF $n(s,0;2v)$ is the number of $s \times 2v$ matrices P of rank s such that $PKP^t = 0$. Write

$$P = \begin{pmatrix} v_1 \\ v_2 \\ \vdots \\ v_s \end{pmatrix},$$

where $v_1, v_2, \cdots, v_s$ are the s row vectors of P. Then

$$v_iKv_j^t = 0 \qquad (1 \leqslant i,j \leqslant s).$$

Since any vector of $\mathbb{F}_q^{(2v)}$ is isotropic, we may choose v_1 to be any nonzero vector of $\mathbb{F}_q^{(2v)}$. Thus, there are $q^{2v} - 1$ possible choices of v_1. Once v_1 is chosen, we may choose v_2 to be any vector in the dual subspace $< v_1 >^\perp$ of the 1-dimensional subspace $< v_1 >$ spanned by v_1 such that v_1 and v_2 are linearly independent. Since $v_1 \in < v_1 >^\perp$, there are $q^{2v-1} - q$ possible choices of v_2. Once v_1 and v_2 are chosen, they span a 2-dimensional totally isotropic subspace $< v_1, v_2 >$. We may choose v_3 to be any vector in the dual subspace $< v_1, v_2 >^{\perp}$ such that v_1, v_2, and v_3 are linearly independent. Since $< v_1, v_2 > \subseteq < v_1, v_2 >^{\perp}$, there are $q^{2v-2} - q^2$ possible choices of v_3. Proceeding in this way, finally when $v_1, v_2, \cdots, v_{s-1}$ are chosen, they span an $(s-1)$-dimensional totally isotropic subspace $< v_1, v_2, \cdots, v_{s-1} >$, and we may choose v_s to be any vector of $< v_1, v_2, \cdots, v_{s-1} >^{\perp}$ not contained in $< v_1, v_2, \cdots, v_{s-1} >$. Thus, there are $q^{2v-s+1} - q^{s-1}$ possible choices of v_s. Therefore,

$$n(s, 0; 2v) = (q^{2v} - 1)(q^{2v-1} - q) \cdots (q^{2v-s+1} - q^{s-1})$$

$$= q^{\frac{s(s-1)}{2}} \prod_{i=v-s+1}^{v} (q^{2i} - 1).$$

□

LEMMA 8.6
Let $s \leqslant v$. Then

$$n(2s, s; 2v) = q^{(2v-s)s} \prod_{i=v-s+1}^{v} (q^{2i} - 1).$$

PROOF $n(2s, s; 2v)$ is the number of $2s \times 2v$ matrices P of rank $2s$ such that

$$PKP^t = \begin{pmatrix} 0 & I^{(s)} \\ -I^{(s)} & 0 \end{pmatrix}.$$

Write

$$P = \begin{pmatrix} P_1 \\ P_2 \end{pmatrix} \begin{matrix} s \\ s \end{matrix}.$$

Then

$$P_1 K P_1^t = 0^{(s)}, \qquad P_1 K P_2^t = I^{(s)},$$

and

$$P_2 K P_2^t = 0^{(s)}.$$

Thus, there are $n(s, 0; 2v)$ possible choices of P_1.

By Lemma 8.4, once P_1 has been chosen, the number of possible choices of P_2 is independent of the particular choice of P_1. Hence, we may choose

$$P_1 = \left(I^{(s)} \; 0^{(s, 2v-s)} \right) = \begin{pmatrix} e_1 \\ e_2 \\ \ddots \\ e_s \end{pmatrix}$$

and then compute the number of possible choices of P_2. Write

$$
P_2 = \begin{pmatrix} v_1 \\ v_2 \\ \cdot \\ \cdot \\ v_s \end{pmatrix},
$$

where $v_1, v_2, \cdots, v_s$ are the s row vectors of P_2. From $P_1 K P_2^t = I^{(s)}$ we deduce that

$$
e_1 K v_1^t = 1, \ e_2 K v_1^t = e_3 K v_1^t = \cdots = e_s K v_1^t = 0,
$$

then v_1 is necessarily of the form

$$
(x_1, x_2, \cdots, x_v, 1, \underbrace{0, \cdots, 0}_{s-1}, x_{v+s+1}, x_{v+s+2}, \cdots, x_{2v}),
$$

where $x_1, x_2, \cdot, x_v, x_{v+s+1}, \cdots, x_{2v}$ can be arbitrary elements of $\mathbb{F}_q$. Thus there are q^{2v-s} possible choices of v_1.

Again by Lemma 8.4, once P_1 and v_1 have been chosen, the number of possible choices of v_2 is independent of the particular choice of P_1 and v_1. Hence, we may choose P_1 as above and choose $v_1 = e_{v+1}$, and then compute the number of possible choices of v_2. From $P_1 K P_2^t = I^{(s)}$ and $P_2 K P_2^t = 0^{(s)}$ we deduce that

$$
e_1 K v_2^t = 0, \ e_2 K v_2^t = 1, \ e_3 K v_2^t = \cdots = e_s K v_2^t = 0, \ e_{v+1} K v_2^t = 0,
$$

then v_2 is necessarily of the form

$$
(0, x_2, \cdots, x_v, 0, 1, \underbrace{0, \cdots, 0}_{s-2}, x_{v+s+1}, x_{v+s+2}, \cdots, x_{2v}),
$$

where $x_2, \cdots, x_v, x_{v+s+1}, x_{v+s+2}, \cdots, x_{2v}$ can be arbitrary elements of $\mathbb{F}_q$. Thus there are q^{2v-s-1} possible choices of v_2.

Proceeding in this way, once P_1 and $v_1, v_2, \cdots, v_{s-1}$ have been chosen, by Lemma 8.4, the number of possible choices of v_s is independent of the particular choices of P_1 and $v_1, v_2, \cdots, v_{s-1}$. Thus, we may choose P_1 as above and choose $v_1 = e_{v+1}, \ v_2 = e_{v+2}, \ \cdots, \ v_{s-1} = e_{v+s-1}$, and then compute the number of possible choices of v_s. From $P_1 K P_2^t = I^{(s)}$ and $P_2 K P_2^t = 0$ we deduce that

$$
e_1 K v_s^t = e_2 K v_s^t = \cdots = e_{s-1} K v_s^t = 0, \qquad e_s K v_s^t = 1,
$$
$$
e_{v+1} K v_s^t = e_{v+2} K v_s^t = \cdots = e_{v+s-1} K v_s^t = 0,
$$

then v_s is necessarily of the form

$$
(\underbrace{0, 0, \cdots, 0}_{s-1}, x_s, x_{s+1}, \cdots, x_v, \underbrace{0, \cdots, 0}_{s-1}, 1, x_{v+s+1}, x_{v+s+2}, \cdots, x_{2v}),
$$

where x_s, s_{s+1}, $\cdots$, x_v, x_{v+s+1}, x_{v+s+2}, $\cdots$, x_{2v} can be arbitrary elements of $\mathbb{F}_q$. Hence, there are $q^{2v-2s+1}$ possible choices of v_s.

Thus, it follows from Lemmas 8.4 and 8.5 that

$$n(2s, s; 2v) = n(s, 0; 2s)q^{2v-s} \cdot q^{2v-s-1} \cdots q^{2v-2s+1}$$

$$= q^{(2v-s)s} \prod_{i=v-s+1}^{v} (q^{2i} - 1).$$

$\square$

THEOREM 8.5

$$|Sp_{2v}(\mathbb{F}_q)| = q^{v^2} \prod_{i=1}^{v} (q^{2i} - 1).$$

PROOF In fact, $|Sp_{2v}(\mathbb{F}_q)| = n(2v, v; 2v)$. $\square$

LEMMA 8.7

Let $2s \leqslant m \leqslant v + s$. Then

$$n(m, s; 2v) = q^{(2v-s)s + \frac{(m-2s)(m-2s-1)}{2}} \prod_{i=v+s-m+1}^{v} (q^{2i} - 1).$$

PROOF $n(m, s; 2v)$ is the number of $m \times 2v$ matrices P of rank m such that $PKP^t = M(m, s)$. Write

$$P = \begin{pmatrix} P_1 \\ P_2 \end{pmatrix} \begin{matrix} 2s \\ m - 2s \end{matrix},$$

then

$$P_1 K P_1^t = \begin{pmatrix} 0 & I^{(s)} \\ -I^{(s)} & 0 \end{pmatrix}, \qquad P_1 K P_2^t = 0,$$

and

$$P_2 K P_2^t = 0^{(m-2s)}.$$

Thus, there are $n(2s, s; 2v)$ possible choices of P_1. Once P_1 has been chosen, by Lemma 8.4, the number of possible choices of P_2 is independent of the particular choice of P_1. Thus, we may choose

$$P_1 = \begin{pmatrix} \overset{s}{I^{(s)}} & \overset{v-s}{0} & \overset{s}{0} & \overset{v-s}{0} \\ 0 & 0 & I^{(s)} & 0 \end{pmatrix}$$

and then compute the number of possible choices of P_2. From $P_1 K P_2^t = 0$, we deduce that P_2 is necessarily of the form

$$P_2 = \begin{pmatrix} \overset{s}{0} & \overset{v-s}{P_{21}} & \overset{s}{0} & \overset{v-s}{P_{22}} \end{pmatrix} m - 2s.$$

Then from $P_2 K P_2^t = 0$, we deduce that

$$\begin{pmatrix} P_{21} & P_{22} \end{pmatrix} \begin{pmatrix} 0 & I^{(v-s)} \\ -I^{(v-s)} & 0 \end{pmatrix} \begin{pmatrix} P_{21} & P_{22} \end{pmatrix}^t = 0.$$

Consequently, there are $n(m-2s, 0; 2[v-s])$ possible choices of P_2. Thus, the lemma follows from Lemmas 8.5 and 8.6.   □

From Lemmas 8.3 and 8.7, Theorem 8.5, and the equality in Equation (8.11) we have the following theorem.

THEOREM 8.6
Let $2s \leqslant m \leqslant v + s$. *Then the number of subspaces of type* (m, s) *in the* $2v$*-dimensional symplectic space over* $\mathbb{F}_q$ *is given by*

$$N(m, s; 2v) = q^{2s(v+s-m)} \frac{\displaystyle\prod_{i=v+s-m+1}^{v} (q^{2i} - 1)}{\displaystyle\prod_{i=1}^{s}(q^{2i} - 1) \prod_{i=1}^{m-2s} (q^i - 1)}.$$

COROLLARY 8.1
Let $m \leqslant v$. *Then the number of* m*-dimensional totally isotropic subspaces in the* $2v$*-dimensional symplectic space over* $\mathbb{F}_q$ *is*

$$N(m, 0; 2v) = \frac{\displaystyle\prod_{i=v-m+1}^{v} (q^{2i} - 1)}{\displaystyle\prod_{i=1}^{m}(q^i - 1)}.$$

COROLLARY 8.2
Let $s \leqslant v$. *Then the number of* $2s$*-dimensional nonisotropic subspaces in the* $2v$*-dimensional symplectic space over* $\mathbb{F}_q$ *is*

$$N(2s, s; 2v) = q^{2s(v-s)} \frac{\displaystyle\prod_{i=v-s+1}^{v} (q^{2i} - 1)}{\displaystyle\prod_{i=1}^{s}(q^{2i} - 1)}.$$

8.2   A-Codes from Symplectic Spaces

Two schemes of A-codes from symplectic spaces will be presented in this section.

Scheme 1

Let v_0 be a fixed nonzero vector in $Sp_{2v}(\mathbb{F}_q)$ and $1 \leqslant m \leqslant v$. Regard the set of m-dimensional totally isotropic subspaces containing v_0 as the set $\mathscr{S}$ of source states, regard the set of 2-dimensional nonisotropic subspaces containing v_0 as the set $\mathscr{E}$ of encoding rules, and regard the set of subspaces of type $(m + 1, 1)$ containing v_0 and not perpendicular to v_0 as the set $\mathscr{M}$ of messages. Given a source state s (i.e., an m-dimensional totally isotropic subspace containing v_0) and an encoding rule e (i.e., a 2-dimensional nonisotropic subspace containing v_0), the space $< s, e >$, generated by s and e, is of type $(m + 1, 1)$ containing v_0 but not perpendicular to v_0, and is regarded as the message $e(s)$.

LEMMA 8.8

Given a message P and any encoding rule e contained in P, there is one and only one source state s such that $< s, e >= P$. Moreover, s is independent of e.

PROOF   We may assume that $e =< u, v_0 >$ with $uKv_0^t = 1$. Since $e \subset P$ and P is of type $(m+1, 1)$, we may assume that P has a matrix representation

$$P = \begin{pmatrix} v_0 \\ u \\ v_2 \\ \vdots \\ v_m \end{pmatrix} \qquad (8.12)$$

such that

$$PKP^t = \begin{pmatrix} 0 & 1 & \\ 1 & 0 & \\ & & 0^{(m-1)} \end{pmatrix}$$

by using the method in the proof of Theorem 8.1.

Let s be an m-dimensional totally isotropic subspace containing v_0 such that $< s, e >= P$. Let $av_0 + bu + a_2v_2 + \cdots + a_mv_m \in s$, then $(av_0 + bu + a_2v_2 + \cdots + a_mv_m)Kv_0^t = b$, and thus $b = 0$. Hence s is the unique subspace spanned by $v_0, v_2, \cdots, v_m$. Let e' be another encoding rule contained in P and let s' be an m-dimensional totally isotropic subspace containing v_0 such that $< s', e' > = P$. From $s' K v_0^t = 0$ we deduce that $s' \subset < v_0, v_2, \cdots, v_m >$ as we did above; therefore $s' = < v_0, v_1, \cdots, v_m > = s$. $\square$

This lemma proves that Scheme 1 results in a Cartesian A-code. Now we compute the cardinal number of the sets $\mathscr{S}$, $\mathscr{E}$, and $\mathscr{M}$.

LEMMA 8.9
The number of source states in Scheme 1 is

$$k = \frac{\displaystyle\prod_{i=v-m+1}^{v-1} (q^{2i} - 1)}{\displaystyle\prod_{i=1}^{m-1} (q^i - 1)}.$$

PROOF By Corollary 8.1, the number of m-dimensional totally isotropic subspaces is

$$N(m, 0; 2v) = \frac{\displaystyle\prod_{i=v-m+1}^{v} (q^{2i} - 1)}{\displaystyle\prod_{i=1}^{m} (q^i - 1)}.$$

The number of 1-dimensional subspaces is

$$N(1, 0; 2v) = \frac{q^{2v} - 1}{q - 1}, \tag{8.13}$$

and the number of 1-dimensional subspaces contained in an m-dimensional totally isotropic subspace is

$$\frac{q^m - 1}{q - 1}.$$

Define a set

$$I = \{(v, P) \mid v \in P\}$$

where v denotes a 1-dimensional subspace and P denotes an m-dimensional totally isotropic subspace. We compute $|I|$ in two different ways.

First, there are $N(1, 0; 2v)$ ways to choose v. For each chosen v, there are $k = |\mathscr{S}|$ ways to choose P such that $v \in P$. Hence,

$$|I| = \frac{q^{2v} - 1}{q - 1} \cdot k.$$

On the other hand, there are $N(m, 0; 2v)$ ways to choose P. For each choice of P, there are $(q^m - 1)/(q - 1)$ ways to choose v such that $v \in P$. Hence,

$$|I| = N(m, 0; 2v) \cdot \frac{q^m - 1}{q - 1}.$$

Combining these two equations, we see that

$$k = \frac{N(m,0;2v) \cdot \frac{q^{m}-1}{q-1}}{\frac{q^{2v}-1}{q-1}}$$

as desired. ☐

LEMMA 8.10
The number of encoding rules in Scheme 1 is

$$b = q^{2(v-1)}.$$

PROOF   The number of 2-dimensional subspaces containing v_0 is

$$\frac{q^{2v} - q}{q^2 - q} = \frac{q^{2v-1} - 1}{q - 1}$$

and the number of 2-dimensional totally isotropic subspaces containing v_0, which is just the number of 2–dimensional subspaces containing v_0 in the dual $(2v-1)$-dimensional subspace $< v_0 >^{\perp}$, is

$$\frac{q^{2v-2} - 1}{q - 1}.$$

Thus

$$b = \frac{q^{2v-1} - 1}{q - 1} - \frac{q^{2v-2} - 1}{q - 1} = q^{2(v-1)}.$$

☐

LEMMA 8.11
The number of messages in Scheme 1 is

$$v = \frac{\prod_{i=v-m+1}^{v-1} (q^{2i} - 1)}{\prod_{i=1}^{m-1} (q^i - 1)} q^{2(v-m)+m-1}.$$

PROOF   The proof of this lemma will be done in two steps. First, we compute the number of subspaces of type $(m+1,1)$ containing a given 1–dimensional subspace v_0. From this result we subtract the number of spaces of type $(m+1,1)$ containing v_0 and perpendicular to v_0. This number is our desired result.

By Theorem 8.6, the number of subspaces of type $(m+1,1)$ is

$$
N(m+1,1;2v) = \frac{\prod\limits_{i=v-m+1}^{v}(q^{2i}-1)}{(q^2-1)\prod\limits_{i=1}^{m-1}(q^i-1)} q^{2(v-m)}. \tag{8.14}
$$

Recall that the number of 1-dimensional subspaces is given by Equation (8.12). The number of 1–dimensional subspaces contained in a subspace of type $(m+1,1)$ is

$$
\frac{q^{m+1}-1}{q-1}. \tag{8.15}
$$

Using a counting technique similar to the one used in the proof of Lemma 8.9, we obtain that the number of $(m+1,1)$ type subspaces containing a given 1-dimensional subspace is given by the following combination of Equations (8.13), (8.14), and (8.15).

$$
\frac{\prod\limits_{i=v-m+1}^{v}(q^{2i}-1)}{(q^2-1)\prod\limits_{i=1}^{m-1}(q^i-1)} \cdot \frac{q^{2(v-m)}(q^{m+1}-1)}{q^{2v}-1}
$$

$$
= \frac{(q^{m+1}-1)\prod\limits_{i=v-m+1}^{v-1}(q^{2i}-1)}{(q^2-1)\prod\limits_{i=1}^{m-1}(q^i-1)} \cdot q^{2(v-m)}. \tag{8.16}
$$

The number of subspaces of type $(m+1,1)$ containing v_0 and perpendicular to v_0 is equal to the number of subspaces of type $(m+1,1)$ containing v_0 and contained in $v_0^{\perp}$. Because of the transitivity of the symplectic space we may assume that v_0 is the vector

$$
v_0 = (1,0,\cdots,0,0\cdots,0),
$$

and thus,

$$
v_0^{\perp} = \{(x_1,\cdots,x_v,0,x_{v+2},\cdots,x_{2v} \mid x_i \in \mathbb{F}_q\}.
$$

Hence, the above number is equal to the number of subspaces of type $(m,1)$

in $Sp_{2(v-1)}(\mathbb{F}_q)$ which equals

$$N(m,1;2(v-1)) = \frac{\displaystyle\prod_{i=(v-1)+1-(m-1)}^{v-1}(q^{2i}-1)}{(q^2-1)\displaystyle\prod_{i=1}^{m-2}(q^i-1)}q^{2(v-1-(m-1))}$$

$$= \frac{\displaystyle\prod_{i=v-m+1}^{v-1}(q^{2i}-1)}{(q^2-1)\displaystyle\prod_{i=1}^{m-2}(q^i-1)}q^{2(v-m)}. \tag{8.17}$$

Subtracting Equation (8.17) from Equation (8.16) gives

$$v = \frac{\displaystyle\prod_{i=v-m+1}^{v-1}(q^{2i}-1)}{(q^2-1)\displaystyle\prod_{i=1}^{m-2}(q^i-1)}q^{2(v-m)}\left(\frac{q^{m+1}-1}{q^{m-1}-1}-1\right)$$

$$= \frac{\displaystyle\prod_{i=v-m+1}^{v-1}(q^{2i}-1)}{\displaystyle\prod_{i=1}^{m-1}(q^i-1)}q^{2(v-m)+(m-1)},$$

which proves the lemma. ☐

The next step is to compute the probabilities P_0 and P_1 for Scheme 1.

LEMMA 8.12
For any message $m_0 \in \mathcal{M}$, we have $|\mathcal{E}(m_0)| = q^{m-1}$.

PROOF Suppose that the message m_0 has the matrix representation in Equation (8.12). The number of encoding rules $< u', v_0 >$ contained in m_0, where $u' \in < v_0 >^\perp$, is given by the number of 1-dimensional subspaces in $< u, v_2, \cdots, v_m >$, which is $(q^m - 1)/(q - 1)$, minus the number of such 1-dimensional subspaces which are perpendicular to v_0, which is $(q^{m-1}-1)/(q-1)$. Thus,

$$|\mathcal{E}(m_0)| = \frac{q^m-1}{q-1} - \frac{q^{m-1}-1}{q-1} = q^{m-1}.$$

☐

LEMMA 8.13
Let s_1, s_2 be two distinct source states and $m_1 \in \mathcal{M}(s_1)$ and $m_2 \in \mathcal{M}(s_2)$

be two messages. If $|\mathscr{E}(m_1, m_2)| > 0$, *then* $|\mathscr{E}(m_1, m_2)| = q^{r-1}$ *where* $r = \dim(s_1 \cap s_2)$, *with* $1 \leqslant r < m$.

PROOF If $|\mathscr{E}(m_1, m_2)| > 0$, there exists an encoding rule $e = < u, v_0 >$ such that $m_1 = < s_1, e >$, $m_2 = < s_2, e >$, and $uKv_0^t \neq 0$. Assume that $s_1 \cap s_2 = S$ with dim $S = r$. Then we can write

$$s_1 = \begin{pmatrix} S \\ S_{10} \end{pmatrix} \begin{matrix} r \\ m-r \end{matrix} \quad \text{and} \quad s_2 = \begin{pmatrix} S \\ S_{20} \end{pmatrix} \begin{matrix} r \\ m-r \end{matrix},$$

with $S_{10} \cap S_{20} = (0)$, and also

$$m_1 = \begin{pmatrix} S \\ S_{10} \\ u \end{pmatrix} \quad \text{and} \quad m_2 = \begin{pmatrix} S \\ S_{20} \\ u \end{pmatrix}$$

with $< S, u > \subset m_1 \cap m_2$. We will show that $m_1 \cap m_2 = < S, u >$. If there is a vector $w \in m_1 \cap m_2$, such that $w \notin < S, u >$, then

$$w = (a_1, a_2, \cdots, a_r)S + (b_1, b_2, \cdots, b_{m-r})S_{10} + cu$$
$$= (a_1', a_2', \cdots, a_r')S + (b_1', b_2', \cdots, b_{m-r}')S_{20} + c'u,$$

where b_i ($1 \leqslant i \leqslant m-r$) are not all zero, and b_i' ($1 \leqslant i \leqslant m-r$) are also not all zero. Since $uKv_0^t \neq 0$ and all vectors in $S + S_{10} + S_{20}$ are perpendicular to v_0, we have $c = c'$ and

$$(a_1, a_2, \cdots, a_r)S + (b_1, b_2, \cdots, b_{m-r})S_{10}$$
$$= (a_1', a_2', \cdots, a_r')S + (b_1', b_2', \cdots, b_{m-r}')S_{20}.$$

But $s_1 \cap s_2 = S$, therefore,

$$(b_1, b_2, \cdots, b_{m-r})S_{10} = (b_1', b_2', \cdots, b_{m-r}')S_{20} = 0$$

is a contradiction. This proves that $m_1 \cap m_2 = < S, u >$.

Therefore any encoding rule contained in $m_1 \cap m_2$ is necessarily of the form $< v_0, x + u >$ where $x \in S$. Two such encoding rules $< v_0, x_1 + u >$ and $< v_0, x_2 + u >$ with $x_1, x_2 \in S$ coincide if and only if $x_1 - x_2 \in < v_0 >$. Thus there are $q^r / q = q^{r-1}$ encoding rules in $m_1 \cap m_2$. Hence, the lemma is proved.
☐

THEOREM 8.7

Suppose that $m = 2$ and the set $\mathscr{E}$ of encoding rules has a uniform probability distribution, then Scheme 1 results in a perfect Cartesian A-code of Type I with the parameters

$$v = \frac{q^{2(v-1)} - 1}{q - 1} \cdot q^{2v-3}, \quad b = q^{2(v-1)}, \quad k = \frac{q^{2(v-1)} - 1}{q - 1},$$

and

$$P_0 = \frac{1}{q^{2v-3}}, \quad P_1 = \frac{1}{q}.$$

PROOF   If $m = 2$, then dim $(s_1 \cap s_2) = 1$ for any two distinct source states s_1 and s_2, and the pair $(\mathcal{M}, \{\mathcal{M}[e] \mid e \in \mathcal{E}\})$ is a strong partially balanced design (SPBD) $2 - (v, b, k; q, 1, 0)$ by Lemmas 8.12 and 8.13. The theorem follows from Lemmas 8.9, 8.10, 8.11, and Theorem 3.3. $\Box$

Scheme 2

Let $v = 2$. Fix a subspace P_0 of type $(3, 1)$ containing a fixed vector v_0 and assume $P_0^{\perp} \neq < v_0 >$. The source states are the nonisotropic subspaces of dimension 2 contained in P_0 and containing v_0. The encoding rules are the 2-dimensional nonisotropic subspaces not contained in P_0 but containing v_0. Finally, the messages are subspaces of type $(3, 1)$ containing v_0 which intersect P_0 in 2-dimensional nonisotropic subspaces. Given a source state s and an encoding rule e, the subspace $< s, e >$ is clearly a subspace of type $(3, 1)$ containing v_0 which intersects P_0 in the 2-dimensional nonisotropic subspace s and will be regarded as the message into which s is encoded under e.

Because of the transitivity of the subspaces of the same type under $Sp_{2v}(\mathbb{F}_q)$, we may without loss of generality assume that

$$P_0 = < e_1, e_2, e_3 >, \quad v_0 = e_1$$

where

$$e_1 = (1, 0, 0, 0), \quad e_2 = (0, 1, 0, 0), \quad e_3 = (0, 0, 1, 0).$$

LEMMA 8.14
Given any message P, there is one and only one source state encoded to P, and this source state is the intersection of P and P_0.

PROOF   Let $s = P \cap P_0$. Clearly, s contains v_0 and is a source state according to the definition. Let e be any encoding rule contained in P. Thus, we have

$$e \cap P_0 = < v_0 >$$

and

$$e \cap s = e \cap (P \cap P_0) = < v_0 > .$$

By the dimension formula

$$\dim s + \dim e = \dim(s \cup e) + \dim(e \cap s),$$

we find that

$$\dim(s \cup e) = \dim s + \dim e - \dim(e \cap s) = 3.$$

Since $s \subset P$, $e \subset P$ we must have $< s, e >= P$, i.e., s is encoded into P under e. Now if there exists another source state s' such that $P = < s', e' >$, where e' is an encoding rule, then

$$s = P \cap P_0 = (s' \cup e') \cap P_0 = (s' \cap P_0) \cup (e' \cap P_0)$$
$$= s' \cup < v_0 >= s'.$$

Thus s is unique.   ☐

Lemma 8.14 shows that Scheme 2 yields a Cartesian A-code.

LEMMA 8.15
The number of source states in Scheme 2 is

$$k = q.$$

PROOF $k = |\mathscr{S}|$ is the number of 2-dimensional nonisotropic subspaces Q, such that $P_0 \supset Q \supset < e_1 >$. Since $e_1 \in Q$, we can assume that any $u \in Q$, minus a linear multiple of e_1, is of the form

$$(0, *, *, 0).$$

Since Q is nonisotropic, there should be a $u_1 \in Q$, such that $e_1 K u_1^t = 1$. Then

$$u_1 = (0, *, 1, 0)$$

and $Q = < e_1, u_1 >$. There are q possible choices of u_1. The lemma follows. ☐

LEMMA 8.16
The number of encoding rules in Scheme 2 is

$$b = (q-1)q.$$

PROOF   A 1-dimensional subspace not contained in P_0 is generated by a vector of the form

$$\lambda_1 e_1 + \lambda_2 e_2 + \lambda_3 e_3 + e_4$$

where $e_4 = (0, 0, 0, 1)$. It generates a 2-dimensional nonisotropic subspace together with e_1, if and only if

$$(\lambda_1 e_1 + \lambda_2 e_2 + \lambda_3 e_3 + e_4) K e_1^t = \lambda_3 \neq 0.$$

Hence, each key e has a unique representation $e = < e_1, u >$ with $u = \lambda_2 e_2 + \lambda_3 e_3 + e_4$ where $\lambda_3 \neq 0$. There are $q(q-1)$ possible choices of u, thus proving the lemma.
⧫

LEMMA 8.17

The number of messages in Scheme 2 is

$$v = q^2.$$

PROOF By Lemma 8.15, the number of 2-dimensional nonisotropic subspaces contained in P_0 and containing v_0 is q. Let s be such a 2-dimensional nonisotropic subspace. Let X be the number of 3-dimensional subspaces containing s. It is easy to see that the number of 2-dimensional subspaces in the 4-dimensional space $V_4(\mathbb{F}_q)$ is

$$\frac{(q^4-1)(q^4-q)}{(q^2-1)(q^2-q)} = \frac{(q^4-1)(q^3-1)}{(q^2-1)(q-1)}.$$

The number of 3-dimensional subspaces in $V_4(\mathbb{F}_q)$ is

$$\frac{(q^4-1)(q^4-q)(q^4-q^2)}{(q^3-1)(q^3-q)(q^3-q^2)} = \frac{q^4-1}{q-1},$$

and the number of 2-dimensional subspaces in $V_3(\mathbb{F}_q)$ is

$$\frac{(q^3-1)(q^3-q)}{(q^2-1)(q^2-q)} = \frac{q^3-1}{q-1}.$$

Using the similar counting technique as one used in the proof of Lemma 8.9, we can have that

$$\frac{(q^4-1)(q^3-1)}{(q^2-1)(q-1)} \cdot X = \frac{q^4-1}{q-1} \cdot \frac{q^3-1}{q-1}.$$

Hence,

$$X = \frac{q^2-1}{q-1} = q+1.$$

The subspace P_0 is one of these 3-dimensional subspaces containing s, therefore, $|\mathcal{M}(s)| = q$ for all $s \in \mathcal{S}$. Thus we have

$$v = |\mathcal{M}| = q^2.$$

⧫

LEMMA 8.18

For any message m_0, we have $|\mathcal{E}(m_0)| = q-1$.

PROOF   We have seen in the proof of Lemma 8.14 that $s = m_0 \cap P_0$ is a 2-dimensional nonisotropic subspace. Without loss of generality, we may assume that

$$m_0 = < e_1, e_2, e_3 >, \quad s = < e_1, e_3 >, \quad v_0 = e_1.$$

From the proof of Lemma 8.15, we know that there exist q 2-dimensional nonisotropic subspaces $e_u^* = < e_1, (0, u, 1, 0) >$ with $u \in \mathbb{F}_q$ in m_0. Clearly, $e_0^* = s \subset P_0$. If $u \neq 0$, e_u^* is not contained in $s = m_0 \cap P_0$, therefore, it is not contained in P_0 and is an encoding rule contained in m_0. Furthermore, we have $< s, e_u^* > = m_0$ when $u \neq 0$, thus $|\mathscr{E}(m_0)| = q - 1$.   ⬚

LEMMA 8.19
For any two distinct messages m_1 and m_2, we have

$$|\mathscr{E}(m_1, m_2)| = 0 \text{ or } 1.$$

PROOF   By the dimension formula,

$$\dim(m_1 \cap m_2) = \dim m_1 + \dim m_2 - \dim(m_1 \cup m_2) = 3 + 3 - 4 = 2.$$

If there exists an encoding rule $e \in \mathscr{E}(m_1, m_2)$, then we must have $e = m_1 \cap m_2$. Hence, $|\mathscr{E}(m_1, m_2)| = 0$ or 1.   ⬚

Thus we have proved following Theorem 8.8 by Lemmas 8.15, 8.16, 8.17, 8.18, 8.19, and Theorem 3.3.

THEOREM 8.8
Suppose that the set $\mathscr{E}$ of encoding rules has a uniform probability distribution, then Scheme 2 results in a perfect Cartesian A-code with the parameters

$$v = q^2, \quad b = (q-1)q, \quad k = q,$$

and

$$P_0 = \frac{1}{q}, \quad P_1 = \frac{1}{q-1}.$$

8.3   A-Codes from Unitary Spaces

c We can construct Cartesian A-codes from the geometry of other types of classical groups, such as unitary and orthogonal groups over finite fields, by the same technique used in Section 8.2. As an example, a new scheme of A-codes over unitary spaces is presented in this section. Schemes 1 and 2 can also be applied over unitary spaces.

Let $\mathbb{F}_{q^2}$ be the finite field with q^2 elements, where q is a prime power. $\mathbb{F}_{q^2}$ has an involutive automorphism, i.e., an automorphism of order 2,

$$a \longmapsto \bar{a} = a^q$$

and the fixed field of this automorphism is $\mathbb{F}_q$.

Let $n \geqslant 1$ and write $n = 2v + \delta$, where $\delta = 0$ or 1. Let

$$H_0 = \begin{pmatrix} 0 & I^{(v)} \\ I^{(v)} & 0 \end{pmatrix}, \quad \text{if} \quad n = 2v,$$

and

$$H_1 = \begin{pmatrix} 0 & I^{(v)} & \\ I^{(v)} & 0 & \\ & & 1 \end{pmatrix}, \quad \text{if} \quad n = 2v + 1.$$

We use the symbol H_δ, where $\delta = 0$ or 1, to cover these two cases.

Let

$$A = (a_{ij})_{1 \leqslant i \leqslant m, 1 \leqslant j \leqslant n}$$

be an $m \times n$ matrix over $\mathbb{F}_{q^2}$. We use $\overline{A}$ to denote the matrix

$$\overline{A} = (\overline{a_{ij}})_{1 \leqslant i \leqslant m, 1 \leqslant j \leqslant n}$$

obtained from A by applying the automorphism $a \mapsto \bar{a}$ to all the mn elements of A. Two $n \times n$ matrices A and B over $\mathbb{F}_{q^2}$ are said to be cogredient, if there is an $n \times n$ nonsingular matrix Q such that $Q A \overline{Q}^t = B$. An $n \times n$ matrix H over $\mathbb{F}_{q^2}$ is said to be Hermitian if $\overline{H}^t = H$. Clearly, the matrices H_0 and H_1 defined above are Hermitian.

Let P be an m-dimensional subspace of $\mathbb{F}_{q^2}^{(n)}$. We use the same letter P to denote a matrix representation of P, i.e., P is also an $m \times n$ matrix whose rows form a basis of P. Thus $P H_\delta \overline{P}^t$ is an $m \times m$ Hermitian matrix. Assume that $P H_\delta \overline{P}^t$ is of rank r, then P is called a subspace of type (m, r).

The unitary group of degree n over $\mathbb{F}_{q^2}$, denoted by $U_n(\mathbb{F}_{q^2})$, is defined to be the set of matrices

$$U_n(\mathbb{F}_{q^2}) = \{T \in GL_n(\mathbb{F}_{q^2}) \,|\, T H_\delta \overline{T}^t = H_\delta\}$$

with matrix multiplication as its group operation. Let $\mathbb{F}_{q^2}^{(n)}$ be the n-dimensional row vector space over $\mathbb{F}_{q^2}$. There is an action of $U_n(\mathbb{F}_{q^2})$ on $\mathbb{F}_{q^2}^{(n)}$, which is called the n-dimensional unitary space, for each $T \in U_n(\mathbb{F}_{q^2}^{(n)})$ defined as follows:

$$\begin{array}{ccc} \mathbb{F}_{q^2}^{(n)} & \longrightarrow & \mathbb{F}_{q^2}^{(n)} \\ (x_1, x_2, \cdots, x_n) & \longmapsto & (x_1, x_2, \cdots, x_n)T. \end{array}$$

Similar to Theorems 8.3 and 8.4, it can be proved that subspaces of type (m, r) exist in the n-dimensional unitary space if and only if $2r \leqslant 2m \leqslant n + r$, and subspaces of the same type are transitive under the act of $U_n(\mathbb{F}_{q^2})$.

For any subspace P, let

$$P^{\perp} = \{x \in \mathbb{F}_{q^2}^{(n)} \mid xH_{\delta}\bar{v}^t = 0 \ \text{ for all } \ v \in P\},$$

$P^{\perp}$ is called the dual space of P. Clearly, dim $P^{\perp} = n - \dim P$.

Scheme 3

Let $n = 3$ and v_0 be a fixed 1-dimensional subspace of type $(1, 0)$. Take the set of subspaces of type $(2, 2)$ containing v_0 to be the set $\mathscr{S}$ of source states, the set of 1-dimensional subspaces who join with v_0 are subspaces of type $(2, 2)$ to be the set of encoding rules and also to be the set $\mathscr{M}$ of messages. For any source state s and encoding rule e, define $e(s) = s \cap e^{\perp}$.

We must show that $s \cap e^{\perp}$ is a message for any source state s and any encoding rule e. Since v_0 is a 1-dimensional subspace of type $(1, 0)$ and $< v_0, e >$ is a 2-dimensional subspace of type $(2, 2)$, we have $v_0 H_1 \bar{e}^t \neq 0$, i.e., $v_0 \notin e^{\perp}$. It follows that $\dim(s \cup e^{\perp}) = 3$, and

$$\dim(s \cap e^{\perp}) = \dim s + \dim e^{\perp} - \dim(s \cup e^{\perp}) = 1$$

by the dimension formula. Furthermore, $< v_0, s \cap e^{\perp} > = s$ is a subspace of type $(2, 2)$, hence, $s \cap e^{\perp}$ is a message. It also shows that for any message m, $s = < v_0, m >$ is the unique source state which can be encoded into m. Hence, the A-code constructed above is Cartesian.

We can assume without loss of generality that

$$v_0 = (1, 0, 0).$$

LEMMA 8.20

The authentication code constructed in Scheme 3 has the size parameters

$$|\mathscr{S}| = q^2, \qquad |\mathscr{E}| = |\mathscr{M}| = q^4.$$

PROOF Let $u = (u_1, u_2, u_3)$ and $v_0 H_1 \bar{u}^t \neq 0$, then

$$v_0 \begin{pmatrix} 0 & 1 & \\ 1 & 0 & \\ & & 1 \end{pmatrix} \bar{u}^t = \bar{u}_2 \neq 0.$$

Hence, each source state has a unique matrix representation

$$\begin{pmatrix} 1 & 0 & 0 \\ 0 & 1 & u_3 \end{pmatrix},$$

so it follows that

$$|\mathscr{S}| = q^2.$$

Each encoding rule and message has a unique matrix representation

$$(u_1, 1, u_3),$$

hence,

$$|\mathscr{E}| = |\mathscr{M}| = q^4.$$

☐

LEMMA 8.21
For any message m, we have $|\mathscr{E}(m)| = q^2$.

PROOF Let $m = (m_1, 1, m_3)$ and $s = \langle v_0, m \rangle$ be the unique source state which can be encoded into m. An encoding rule $e = (u_1, 1, u_3)$ belongs to $\mathscr{E}(m)$ if and only if $m = s \cap e^\perp$, i.e., if and only if $m \in e^\perp$. Since

$$(m_1, 1, m_3) \begin{pmatrix} 0 & 1 & \\ 1 & 0 & \\ & & 1 \end{pmatrix} (\overline{u}_1, 1, \overline{u}_3)^t = \overline{u}_1 + m_1 + m_3 \overline{u}_3 = 0,$$

it follows that

$$u_1 = -\overline{m}_1 - \overline{m}_3 u_3$$

where u_3 can be any element of $\mathbb{F}_{q^2}$. Hence, we have $|\mathscr{E}(m)| = q^2$.   ☐

LEMMA 8.22
For any two messages m and m', $|\mathscr{E}(m, m')| = 0$ or 1.

PROOF Let $m = (m_1, 1, m_3)$ and $m' = (m'_1, 1, m'_3)$ be two distinct messages. Assume that the encoding rule $e = (u_1, 1, u_3)$ belongs to $\mathscr{E}(m, m')$. Then we have $m_3 \neq m'_3$, otherwise $s = \langle v_0, m \rangle = \langle v_0, m' \rangle$ and $m = m' = s \cap e^\perp$, which is a contradiction. As shown in the proof of Lemma 8.21, we have

$$mH_1\overline{e}^t = \overline{u}_1 + m_1 + m_3\overline{u}_3 = 0,$$
$$m'H_1\overline{e}^t = \overline{u}_1 + m'_1 + m'_3\overline{u}_3 = 0,$$

and the system of linear equations determines (u_1, u_3) uniquely. The lemma is proved.   ☐

We can deduce the following theorem by Lemmas 8.20, 8.21, and 8.22, and Theorem 3.3.

THEOREM 8.9
Suppose that the set of encoding rules has a uniform probability distribution, then Scheme 3 results in a perfect Cartesian A-code of Type I with the parameters

$$k = q^2, \qquad b = v = q^4,$$

and

$$P_0 = \frac{1}{q^2}, \qquad P_1 = \frac{1}{q^2}.$$

8.4 Comments

For the convenience of readers, we present the background of symplectic spaces over finite fields in Section 8.1. For further understanding of finite geometries, readers can refer to Wan [52]. The text pp. 153-168 is cited from the text pp. 107-127 of [52] with kind permission of Science Press.

Scheme 1 and Scheme 2 (pp. 169-172) are cited from the text pp. 921-922 of Wan et al. [53] with kind permission of IEEE Press, and Scheme 3 is due to Feng and Wan [11]. For simplicity, only the case of perfect authentication codes constructed by these schemes is considered in this chapter.

8.5 Exercises

8.1 Let v_0 be a fixed 1-dimensional subspace of type $(1, 0)$ in the unitary space $U_4(\mathbb{F}_{q^2})$. Take the set of subspaces of type $(2, 0)$ containing v_0 to be the set $\mathscr{S}$ of source states, the set of subspaces of type $(2, 2)$ containing v_0 to be the set $\mathscr{E}$ of encoding rules, and the set of subspaces of type $(3, 2)$ containing v_0 and not perpendicular to v_0 to be the set $\mathscr{M}$ of messages. For any source state s and any encoding rule e, define $e(s) = <s, e>$. Prove that the constructed A-code is perfect Cartesian with parameters $v = q^2(q^2 + 1)$, $b = q^4$, $k = q^2 + 1$, and $P_0 = P_1 = q^{-2}$ when $\mathscr{E}$ has a uniform probability distribution.

8.2 Let v_0 be a fixed 1-dimensional subspace of type $(1, 0)$ in the unitary space $U_4(\mathbb{F}_{q^2})$ and let P_0 be a fixed 3-dimensional subspace of type $(3, 2)$ containing v_0 and $P_0 \neq v_0^{\perp}$. Take the set of subspaces of type $(2, 2)$ contained in P_0 and containing v_0 to be the set $\mathscr{S}$ of source states, the set of subspaces of type $(2, 2)$ not contained in P_0 but containing v_0 to be the set $\mathscr{E}$ of encoding rules, and the set of subspaces of type $(3, 2)$

containing v_0 which intersect P_0 in subspaces of type $(2,2)$ to be the set $\mathcal{M}$ of messages. For any source state s and any encoding rules e, define $e(s) = < s, e >$. Prove that the constructed A-code is perfect Cartesian with parameters $v = q^4$, $b = q^2(q^2 - 1)$, $k = q^2$ and $P_0 = q^{-1}$, $P_1 = (q^2 - 1)^{-1}$, when $\mathcal{E}$ has a uniform probability distribution.

8.3 Let $v \geqslant 2$ and v_0 be a fixed vector in the symplectic space $Sp_{2v}(\mathbb{F}_q)$. Take the set of subspaces of type $(2,1)$ containing v_0 to be the set $\mathcal{S}$ of source states, the set of 1-dimensional subspaces whose joins with v_0 are subspaces of type $(2,1)$ to be the set $\mathcal{E}$ of encoding rules and also the set $\mathcal{M}$ of messages. For any source state s and encoding rule e, define $e(s) = s \cap e^{\perp}$. Prove that the constructed A-code is Cartesian with parameters $k = q^{2v-2}$, $b = v = q^{2v-1}$, and $P_0 = P_1 = q^{-1}$ when $\mathcal{E}$ has a uniform probability distribution.

Chapter 9

Authentication/Secrecy Schemes

Two of the main applications of cryptography are the provisions of secrecy and authentication for messages. Consider the model with three parties. A transmitter communicates a sequence of distinct source states from a set $\mathscr{S}$ to a receiver by encoding them using one from a set of encoding rules $\mathscr{E}$. Each encoding rule is an injective mapping from $\mathscr{S}$ to the set of messages $\mathscr{M}$. The receiver recovers the source states from the received messages by determining their (unique) preimages under the agreed encoding rule. The receiver accepts a message as authentic if it lies in the image of the agreed encoding rule. An opponent observes the resulting sequence of messages and attempts to determine the information about the corresponding source states, thereby compromising their secrecy, or attempts to determine another message which will be accepted by the receiver as authentic, thereby deceiving the receiver. The encoding rule may be regarded as the cryptographic transformation corresponding to a secret key.

In 1949 Shannon [40] showed how to construct systems offering unconditional secrecy, that is, theoretically perfect-secrecy systems. Following this work on secrecy, Simmons [41] and others ([12], [6]) have considered systems which offer unconditional authentication. Both unconditional secrecy and unconditional authentication are achieved at the expense of requiring a very large number of keys. To reduce the number of required keys and hence to reduce the key size in these systems as much as possible is an essential point to be considered.

In Section 9.1 several definitions of perfect secrecy are introduced. Lower bounds are given for the number of keys in such perfect systems. Theorems characterizing systems meeting these lower bounds (key minimal systems) are obtained. The same subjects for the authentication systems with some perfect secrecy are discussed in Section 9.3. Some perfect secrecy schemes without authentication are constructed in Section 9.2. Authentication/secrecy schemes with key-entropy minimal and perfect secrecy are constructed in Section 9.4.

9.1   Perfect Secrecy Schemes

Let us recall some notations and assumptions first. Let $\mathscr{S}$, $\mathscr{M}$, and $\mathscr{E}$ be the set of source states, the set of encoded message and the set of encoding rules, respectively, while S, M, and E denote the random variable of source states, of messages, and of encoding rules, respectively. The probability distribution of S and E is denoted by p_S and p_E, respectively. The probability distribution of M is determined by p_S and p_E. For simplicity, we write $p(s)$ to denote the probability for occurrence of source state $s \in \mathscr{S}$, similarly for $p(e)$ ($e \in \mathscr{E}$) and $p(m)$ ($m \in \mathscr{M}$). We also abbreviate the notation of conditional probability, for example, we write $p(e|m)$ instead of $p(E = e|M = m)$.

We assume that p_S and p_E are independent. This means

$$p(s, e) = p(S = s, E = e) = p(s)p(e).$$

For any message $m \in \mathscr{M}$, we have

$$p(m) = \sum p(s)p(e)$$

where the sum is over all pairs (s, e) such that $e(s) = m$.

We require that for every encoding rule e

$$p(e) > 0,$$

for every source state s

$$p(s) > 0,$$

and for every message m

$$p(m) > 0,$$

otherwise those encoding rules, source states, and messages with probability zero of occurrence can be removed from the sets under consideration.

We consider only the case that the transmitter sends a sequence of distinct source states by the same encoding rule. It means that the transmitter never sends the same source state twice by the same encoding rule. This also implies that the transmitter never uses the same message twice under the same encoding rule. Otherwise, we can not resist the double sending attack (the opponent repeats a message, sent by the transmitter, to the receiver). Hence, we assume that, for any set of source states $s^r = (s_1, s_2, \cdots s_r)$, its probability of occurrence is positive only if the r source states are distinct.

Suppose that the opponent has observed a set $m_1, m_2, \cdots, m_r$ of r messages sent by the transmitter using the same encoding rule, then

(i) the opponent may gain some information about the set s^r of r source states corresponding to the set $m_1, m_2, \cdots, m_r$;

(ii) besides the information in (i), the opponent may even gain some information about the possible orderings of source states in s^r.

Consider the above two kinds of attacks from the opponent. For any r source states, or any r messages, we will distinguish two cases: ordered or unordered. So far in the previous chapters what we have considered is only the ordered case.

Let $s^r = (s_1, s_2, \cdots, s_r)$ be an r-tuple (ordered) of distinct source states, and let $\mathscr{S}^r$ be the set of all r-tuples of distinct source states. Denote the random variable taking its values in $\mathscr{S}^r$ by S^r. As we did in Equation (2.1), assume that S^r has a positive distribution, which means that $p(s^r) > 0$ for any $s^r \in \mathscr{S}^r$.

We have already defined the notation $\mathscr{E}(m^r)$ for an r-tuple $m^r = (m_1, m_2, \cdots, m_r)$ of distinct messages and the notation $\overline{\mathscr{M}^r}$ for the set of all allowable r-tuples of distinct messages in Section 2.1. The random variable M^r takes its values in the set $\overline{\mathscr{M}^r}$; we have, for any $m^r \in \overline{\mathscr{M}^r}$,

$$p(m^r) = \sum_{e \in \mathscr{E}(m^r)} p(e)p\big(f_e(m^r)\big) > 0,$$

where $f_e(m^r) \in \mathscr{S}^r$ is the preimage of m^r under the encoding rule e.

Let s'_r denote an r-subset (unordered) of distinct elements in $\mathscr{S}$ and let $\mathscr{S}_r$ be the set of all r-subsets of distinct elements of $\mathscr{S}$. By S_r we denote the random variable taking its values in $\mathscr{S}_r$. Furthermore, we assume that $\mathscr{S}_r$ has a positive distribution.

Let m'_r be an r-subset of distinct messages, and define

$$\mathscr{E}(m'_r) = \{e \in \mathscr{E} \mid m'_r \subset e(\mathscr{S})\}$$

and

$$\overline{\mathscr{M}_r} = \{m'_r \mid \mathscr{E}(m'_r) \neq \emptyset\}.$$

Each element m'_r of $\overline{\mathscr{M}_r}$ is said to be allowable. This means that there exists an encoding rule e and an r-subset $s'_r \in \mathscr{S}^r$ such that

$$e(s'_r) = m'_r.$$

Let M_r denote the random variable taking its values in the set $\overline{\mathscr{M}_r}$. We have, for any $m'_r \in \overline{\mathscr{M}_r}$,

$$p(m'_r) = \sum_{e \in \mathscr{E}(m'_r)} p(e)p\big(f_e(m'_r)\big) > 0,$$

where $f_e(m'_r) \in \mathscr{S}_r$ is the preimage of m'_r under the encoding rule e. We refer to $\mathscr{A} = (\mathscr{S}, \mathscr{M}, \mathscr{E}, p_S, p_E)$ as a secrecy scheme when its secrecy only is considered. Denote $|\mathscr{S}| = k$, $|\mathscr{M}| = v$, $|\mathscr{E}| = b$ as before.

We introduce two definitions of perfect secrecy.

DEFINITION 9.1   *Given $t \geqslant 1$, a secrecy scheme $\mathscr{A} = (\mathscr{S}, \mathscr{M}, \mathscr{E}, p_S, p_E)$ is said to provide Unordered Perfect t-fold secrecy $\{U(t)$-secrecy$\}$ if, for every* allowable t-subset m_t' of $\mathcal{M}$ and for every t-subset s_t' of distinct source states from $\mathcal{S}$,

$$p(s_t'|m_t') = p(s_t').$$

DEFINITION 9.2   *Given* $t \geq 1$, *a secrecy scheme* $\mathcal{A} = (\mathcal{S}, \mathcal{M}, \mathcal{E}, p_S, p_E)$ *is said to provide Ordered Perfect t-fold secrecy $\{O(t)$-secrecy$\}$ if, for every allowable t-tuple $m^t = (m_1, m_2, \cdots, m_t)$ of distinct messages from $\mathcal{M}$ and for every t-tuple $s^t = (s_1, s_2, \cdots, s_t)$ of distinct source states from $\mathcal{S}$,*

$$p(s^t|m^t) = p(s^t).$$

LEMMA 9.1

If $\mathcal{A} = (\mathcal{S}, \mathcal{M}, \mathcal{E}, p_S, p_E)$ *provides $O(t)$-secrecy $(1, \leq t < k)$, then it also provides $O(r)$-secrecy for every r satisfying $1 \leq r \leq t$.*

PROOF   Suppose that s^{t-1} is a $(t-1)$-tuple of distinct source states and m^{t-1} is an allowable $(t-1)$-tuple of distinct messages. Hence, we have $p(m^{t-1}) > 0$. Let $X(s^{t-1})$ be the set of all t-tuples of distinct source states which agree with s^{t-1} in the first $(t-1)$ positions. Similarly, let $X(m^{t-1})$ be the set of all allowable t-tuples of distinct messages which agree with m^{t-1} in the first $(t-1)$ positions. Then

$$
\begin{aligned}
p(s^{t-1}|m^{t-1}) &= \sum_{s^t \in X(s^{t-1})} p(s^t|m^{t-1}) \\
&= \sum_{s^t \in X(s^{t-1})} p(m^{t-1}|s^t)p(s^t)/p(m^{t-1}) \\
&= \sum_{s^t \in X(s^{t-1})} \left(\sum_{m^t \in X(m^{t-1})} p(m^t|s^t)p(s^t) \right) \Big/ \left(\sum_{m^t \in X(m^{t-1})} p(m^t) \right) \\
&= \sum_{s^t \in X(s^{t-1})} \left(\sum_{m^t \in X(m^{t-1})} p(m^t)p(s^t) \right) \Big/ \left(\sum_{m^t \in X(m^{t-1})} p(m^t) \right) \\
&= \sum_{s^t \in X(s^{t-1})} p(s^t) \\
&= p(s^{t-1}),
\end{aligned}
$$

which means that $\mathcal{A}$ provides $O(t-1)$-secrecy. The result then follows by induction. $\square$

LEMMA 9.2

If $\mathcal{A} = (\mathcal{S}, \mathcal{M}, \mathcal{E}, p_S, p_E)$ *provides $O(t)$-secrecy $(1 \leq t < k)$, then it also provides $U(r)$-secrecy for every r satisfying $1 \leq r \leq t$.*

PROOF   Suppose that m'_r is any element of $\overline{\mathcal{M}_r}$ and s'_r is any element of $\mathcal{S}_r$. Hence, we have $p(m'_r) > 0$. Let $T(m'_r)$ denote the subset of $\overline{\mathcal{M}^t}$ consisting of those t-tuples containing all the elements of m'_r. Similarly, let $T(s'_r)$ denote the set of those t-tuples of distinct source states which contain all elements of s'_r. Then

$$
\begin{aligned}
p(s'_r|m'_r) &= \sum_{s^t \in T(s'_r)} p(s^t|m'_r) \\
&= \sum_{s^t \in T(s'_r)} p(s^t, m'_r)/p(m'_r) \\
&= \sum_{s^t \in T(s'_r)} \left(\sum_{m^t \in T(m'_r)} p(m^t)p(s^t|m^t) \right) \Big/ \left(\sum_{m^t \in T(m'_r)} p(m^t) \right) \\
&= \sum_{s^t \in T(s'_r)} p(s^t) \sum_{m^t \in T(m'_r)} p(m^t) \Big/ \left(\sum_{m^t \in T(m'_r)} p(m^t) \right) \\
&= p(s'_r).
\end{aligned}
$$

The result follows. □

$\mathcal{A}$ is said to provide Stinson Perfect t-fold secrecy $\{S(t)$-secrecy$\}$ if $\mathcal{A}$ provides $U(r)$-secrecy for every $1 \leqslant r \leqslant t$. Lemma 9.2 shows that $O(t)$-secrecy is stronger than $U(t)$-secrecy and $S(t)$-secrecy.

LEMMA 9.3
If $\mathcal{A}$ provides $U(t)$-secrecy, then for every allowable t-subset of distinct messages m'_t and for every t-subset of distinct source states s'_t there exists an encoding rule e such that

$$
e(s'_t) = m'_t.
$$

PROOF   Suppose there exists a pair of t-subsets s'_t, m'_t (allowable) such that there is no encoding rule which maps s'_t onto m'_t. Then, clearly

$$
p(s'_t \mid m'_t) = 0
$$

which contradicts the assumption of $U(t)$-secrecy since

$$
p(s'_t) > 0
$$

by assumption that $\mathcal{S}^t$ has a positive probability distribution. □

We now study the bound of the number of encoding rules for secrecy systems which provide $U(t)$-secrecy or $O(t)$-secrecy.

LEMMA 9.4

If $\mathscr{A}$ provides $U(t)$-secrecy, then

$$b \geqslant |\overline{\mathscr{M}_t}|,$$

where $\overline{\mathscr{M}_t}$ is the set of allowable t-subsets of $\mathscr{M}$. Moreover, if

$$b = |\overline{\mathscr{M}_t}|,$$

then:

(i) *If s'_t is any element of $\mathscr{S}_t$ and m'_t is any element of $\overline{\mathscr{M}_t}$, there exists a unique encoding rule e such that*

$$e(s'_t) = m'_t.$$

(ii) *For every encoding rule e, if m'_t is any element of $\overline{\mathscr{M}_t}$ with $m'_t \subset e(\mathscr{S})$, then*

$$p(e) = p(m'_t).$$

PROOF   Let s'_t be a t-subset of $\mathscr{S}_t$. Then, by Lemma 9.3, if m'_t is any element of $\overline{\mathscr{M}_t}$, there exists an encoding rule e with

$$e(s'_t) = m'_t.$$

Therefore, if we fix s'_t and let m'_t range over all elements of $\overline{\mathscr{M}_t}$, we obtain $|\overline{\mathscr{M}_t}|$ different encoding rules. The bound follows.

Now suppose that $b = |\overline{\mathscr{M}_t}|$. Following the above argument, it is clear that fixing s'_t and letting m'_t range over all the elements of $\overline{\mathscr{M}_t}$ exhausts the set of encoding rules. Statement (i) then follows immediately.

To establish (ii), suppose that s'_t is any element of $\mathscr{S}_t$ and m'_t is any element of $\overline{\mathscr{M}_t}$. In addition let e be the unique encoding rule which maps s'_t onto m'_t. Then

$$p(s'_t|m'_t) = p(s'_t, m'_t)/p(m'_t) = p(s'_t)p(e)/p(m'_t).$$

But, by the definition of $U(t)$-secrecy,

$$p(s'_t|m'_t) = p(s'_t),$$

hence

$$p(e) = p(m'_t)$$

and (ii) follows.   ☐

THEOREM 9.1

If $\mathscr{A}$ provides $U(t)$-secrecy, then

$$b \geqslant (v/k)\binom{k}{t}.$$

Moreover, if

$$b = (v/k)\binom{k}{t},$$

then

(i) *If $t > 1$, for any pair of encoding rules e_1, e_2, either*

$$e_1(\mathscr{S}) = e_2(\mathscr{S})$$

or $e_1(\mathscr{S})$ and $e_2(\mathscr{S})$ are disjoint.

(ii) *If e_1 and e_2 are encoding rules satisfying*

$$e_1(\mathscr{S}) = e_2(\mathscr{S}),$$

then

$$p(e_1) = p(e_2) = p(m_t')$$

for every $m_t' \subset e_1(\mathscr{S})$.

PROOF By Lemma 9.4, to establish the bound we need only to show that

$$|\overline{\mathscr{M}_t}| \geqslant (v/k) \cdot \binom{k}{t}.$$

Choose a message m, then, since we assume throughout that $p(m) > 0$, there exists a source state s and an encoding rule e with

$$e(s) = m.$$

It is clear that

$$|e(\mathscr{S})| = k,$$

and hence, there are at least $\binom{k-1}{t-1}$ allowable t-subsets of $\mathscr{M}$ which includes m. Since there are precisely v choices for m this gives us a total of $v \cdot \binom{k-1}{t-1}$ (not necessarily distinct) allowable t-subsets of $\mathscr{M}$. Each such allowable t-subset has been counted t times, which implies

$$|\overline{\mathscr{M}_t}| \geqslant v \cdot \binom{k-1}{t-1} \Big/ t = (v/k) \cdot \binom{k}{t}.$$

The bound follows.

Now suppose that

$$b = (v/k) \cdot \binom{k}{t},$$

and hence,

$$b = |\overline{\mathscr{M}_t}| = v \cdot \binom{k-1}{t-1} \Big/ t.$$

Suppose that $t > 1$. We know that each message m is included at least in $\binom{k-1}{t-1}$ allowable t-subsets, since there are only $(v/t) \cdot \binom{k-1}{t-1}$ allowable t-subsets in total, so each message m is contained in precisely $\binom{k-1}{t-1}$ allowable t-subsets. Now if e is any encoding rule for which $m \in e(\mathscr{S})$, then since $|e(\mathscr{S})| = k$, e itself will immediately yield $\binom{k-1}{t-1}$ allowable t-subsets containing m. Hence, if e' is any other encoding rule for which $m \in e'(\mathscr{S})$, then $e(\mathscr{S}) = e'(\mathscr{S})$ since there are no more allowable t-subsets containing m. The above applies for all messages m and (i) follows.

Statement (ii) is immediate from application of Lemma 9.4 (ii).   ▯

Theorem 9.1 (i) does not hold for the case $t = 1$. Counterexamples will been given later.

We can get a result as Theorem 9.1 for $O(t)$-secrecy.

LEMMA 9.5

If $\mathscr{A}$ provides $O(t)$-secrecy and r satisfies $1 \leqslant r \leqslant t$, then for every allowable r-tuple of distinct messages m^r and for every r-tuple of distinct source states s^r there exists an encoding rule e such that

$$e(s^r) = m^r.$$

PROOF   Suppose there exists a pair of r-tuples s^r, m^r (allowable) such that there is no encoding rule which maps s^r to m^r. Then, clearly

$$p(s^r | m^r) = 0$$

which contradicts the assumption of $O(r)$-secrecy since

$$p(s^r) > 0$$

by the assumption that $\mathscr{S}^r$ has a positive probability distribution.   ▯

LEMMA 9.6

If $\mathscr{A}$ provides $O(t)$-secrecy, then

$$b \geqslant |\overline{\mathscr{M}^t}|,$$

where $\overline{\mathscr{M}^t}$ is the set of allowable t-tuples of distinct messages of $\mathscr{M}$. Moreover, if

$$b = |\overline{\mathscr{M}^t}|,$$

then:

(i) *If s^t is any t-tuple of distinct elements of $\mathscr{S}$ and m^t is any element of $\overline{\mathscr{M}^t}$, there exists a unique encoding rule e such that*

$$e(s^t) = m^t.$$

(ii) *For every encoding rule* e, *if* $m^t \subset e(\mathscr{S})$, *then*

$$p(e) = p(m^t).$$

PROOF Let s^t be a t-tuple of distinct elements of $\mathscr{S}$. Then, by Lemma 9.5, if m^t is any element of $\overline{\mathscr{M}^t}$, there exists an encoding rule e with

$$e(s^t) = m^t.$$

Therefore, if we fix s^t and let m^t range over all elements of $\overline{\mathscr{M}^t}$, we obtain $|\overline{\mathscr{M}^t}|$ different encoding rules. The bound follows.

Now suppose $b = |\overline{\mathscr{M}^t}|$. Following the above argument, it is clear that fixing s^t and letting m^t range over all the elements of $\overline{\mathscr{M}^t}$ exhausts the set of encoding rules. Statement (i) then follows immediately.

To establish (ii) suppose s^t is any t-tuple of distinct source states and let e be the unique encoding rule which maps s^t onto m^t. Then

$$p(s^t|m^t) = p(s^t, m^t)/p(m^t) = p(e)p(s^t)/p(m^t),$$

but, by definition of $O(t)$-secrecy,

$$p(s^t|m^t) = p(s^t).$$

Hence,

$$p(e) = p(m^t).$$

Statement (ii) follows immediately. ⬜

THEOREM 9.2
If $\mathscr{A}$ *provides* $O(t)$-*secrecy, then*

$$b \geqslant v \cdot (k-1)!/(k-t)!.$$

Moreover, if

$$b = v \cdot (k-1)!/(k-t)!,$$

then:

(i) *If* $t > 1$, *for any pair of encoding rules* e_1, e_2, *either*

$$e_1(\mathscr{S}) = e_2(\mathscr{S})$$

or $e_1(\mathscr{S})$ *and* $e_2(\mathscr{S})$ *are disjoint.*

(ii) *If* e_1 *and* e_2 *are encoding rules satisfying*

$$e_1(\mathscr{S}) = e_2(\mathscr{S}),$$

then

$$p(e_1) = p(e_2) = p(m^t),$$

for any $m^t \in \overline{\mathscr{M}^t}$ *with* $m^t \subset e_1(\mathscr{S})$.

PROOF   By Lemma 9.6, to establish the bound we only need to show that

$$|\overline{\mathcal{M}^t}| \geqslant v \cdot (k-1)!/(k-t)!.$$

Choose any message m, then, since $p(m) > 0$ there exists a source state s and an encoding rule e with

$$e(s) = m.$$

Now, it is clear that

$$|e(\mathcal{S})| = k,$$

and hence there are at least $(k-1)!/(k-t)!$ allowable t-tuples of distinct elements of $\mathcal{M}$ which have m as their first entry. Since there are precisely v choices for m, this gives us

$$|\overline{\mathcal{M}^t}| \geqslant v \cdot (k-1)!/(k-t)!$$

and the bound follows.

Now suppose that

$$b = v \cdot (k-1)!/(k-t)!,$$

and hence,

$$b = |\overline{\mathcal{M}^t}| = v \cdot (k-1)!/(k-t)!.$$

We know that each message m is included as the first element in at least $(k-1)!/(k-t)!$ allowable t-tuples; since there are totally $v \cdot (k-1)!/(k-t)!$ allowable t-tuples, each message m is contained as the first element in precisely $(k-1)!/(k-t)!$ allowable t-tuples.

Suppose that $t > 1$. Now, if e is an encoding rule for which $m \in e(\mathcal{S})$, since $|e(\mathcal{S})| = k$, e itself will immediately yield $(k-1)!/(k-t)!$ allowable t-tuples with first element m. If e' is any other encoding rule for which $m \in e'(\mathcal{S})$, then $e(\mathcal{S}) = e'(\mathcal{S})$ since there are no more allowable t-tuples with m as the first element. This argument applies for all messages m, and (i) follows.

Statement (ii) follows immediately from Lemma 9.6 (ii). $\square$

Theorem 9.2 (i) does not hold for $t = 1$. Counterexamples will be given later.

9.2   Construction of Perfect Secrecy Schemes

It is of interest to construct secrecy systems for which the numbers of encoding rules meet the lower bounds established in Section 9.1, since it is always desirable to minimize the key size. We divide our construction into two categories, namely, those satisfying the bounds of Theorems 9.1 and 9.2, respectively. We call this type of system as perfect.

In this section we study systems with secrecy only, as in the above section. When no authentication is required, it is reasonable to assume that $v = k$, since information redundancy is not absolutely necessary in this case. Thus, the lower bound of the number of encoding rules established in Theorem 9.1 is $b = \binom{k}{t}$. We may identify $\mathscr{S}$ with $\mathscr{M}$; each encoding rule is then no more than a permutation on $\mathscr{M}$. If $b = \binom{k}{t}$, by Theorem 9.1 (ii), each encoding rule must be equiprobable. These constraints now enable us to give a purely combinatorially necessary and sufficient condition for a set of $\binom{k}{t}$ permutations on $\mathscr{M}$ to form a system providing $U(t)$-secrecy. Obviously, each t-subset of $\mathscr{M}$ is allowable.

THEOREM 9.3
The scheme $\mathscr{A} = (\mathscr{S}, \mathscr{M}, \mathscr{E}, p_S, p_E)$ *with* $|\mathscr{E}| = \binom{k}{t}$ *provides* $U(t)$-*secrecy if and only if*

(i) $p(e) = 1/\binom{k}{t}$ *for every encoding rule* $e \in \mathscr{E}$.

(ii) *For every pair of t-subsets s'_t, m'_t of $\mathscr{M}$, there exists a unique encoding rule e with $e(s'_t) = m'_t$.*

PROOF   Suppose that $\mathscr{A}$ provides $U(t)$-secrecy. Then (i) holds by Theorem 9.1 (ii). Moreover, (ii) holds by Lemma 9.4 (i).

Conversely, now suppose that (i) and (ii) hold and s'_t is a t-subset of $\mathscr{S}$ $(= \mathscr{M})$, and m'_t is a t-subset of $\mathscr{M}$ (and hence allowable) with unique encoding rule e such that $e(s'_t) = m'_t$. Then

$$p(s'_t|m'_t) = p(s'_t, m'_t)/p(m'_t) = p(s'_t)p(e)/p(m'_t) = p(s'_t)\bigg/\left(\binom{k}{t}p(m'_t)\right).$$

But, by definition

$$p(m'_t) = \sum_{\{(e, s'_t)|e(s'_t)=m'_t\}} p(s'_t)p(e) = \sum_{s'_t \in \mathscr{S}} p(s'_t)\bigg/\binom{k}{t} = 1\bigg/\binom{k}{t}.$$

Substituting the second equality into the first one, hence, as required,

$$p(s'_t|m'_t) = p(s'_t)$$

and the result follows.   ⬜

THEOREM 9.4
Suppose $\mathscr{A} = (\mathscr{S}, \mathscr{M}, \mathscr{E}, p_S, p_E)$, *with* $|\mathscr{M}| = k$ *and* $|\mathscr{E}| = \binom{k}{t}$ *provides* $U(t)$-*secrecy. Then for every $r \leqslant t$, $\mathscr{A}$ also provides $U(r)$-secrecy if and only if, for every pair of r-subsets s'_r, m'_r of $\mathscr{M}$, there exist precisely w' encoding rules e with*

$$e(s'_r) = m'_r,$$

where

$$w' = \binom{k}{t} \Big/ \binom{k}{r}.$$

PROOF   Let $\mathscr{E}(s'_r, m'_r)$ denote the set of encoding rules which map s'_r to m'_r. By definition, $\mathscr{A}$ provides $U(r)$-secrecy

$$\text{if and only if} \qquad p(s'_r|m'_r) = p(s'_r).$$

This holds

$$\text{if and only if} \qquad p(m'_r|s'_r) = p(m'_r),$$

$$\text{if and only if} \qquad \sum_{e \in \mathscr{E}(s'_r, m'_r)} p(e) = p(m'_r), \tag{9.1}$$

$$\text{if and only if} \qquad |\mathscr{E}(s'_r, m'_r)| \Big/ \binom{k}{t} = p(m'_r).$$

First, suppose Equation (9.1) holds. Now if we fix m'_r and let s'_r range over all $\binom{k}{r}$ possible r-subsets of $\mathscr{M}$, then the sets $\mathscr{E}(s'_r, m'_r)$ will be pairwise disjoint and their union is $\mathscr{E}$. Moreover, since the right-hand side of Equation (9.1) is fixed, they must all have the same size. Hence,

$$|\mathscr{E}(s'_r, m'_r)| = |\mathscr{E}| \Big/ \binom{k}{r} = \binom{k}{t} \Big/ \binom{k}{r} \tag{9.2}$$

as required.

Now, suppose that Equation (9.2) holds for all s'_r and m'_r. By definition,

$$p(m'_r) = \sum_{s'_r \in \mathscr{S}_r} p(s'_r) \sum_{e \in \mathscr{E}(s'_r, m'_r)} p(e)$$

$$= \sum_{s'_r \in \mathscr{S}_r} p(s'_r)|\mathscr{E}(s'_r, m'_r)| \Big/ \binom{k}{t}$$

$$= \sum_{s'_r \in \mathscr{S}_r} p(s'_r) \cdot 1 \Big/ \binom{k}{r} = 1 \Big/ \binom{k}{r} = |\mathscr{E}(s'_r, m'_r)| \Big/ \binom{k}{t}.$$

Equation (9.1) follows. ⬜

DEFINITION 9.3   *Suppose $\mathscr{E}$ is a set of permutations on the set $\mathscr{S}$. Then $\mathscr{E}$ is said to be $\{t, w\}$-homogeneous on $\mathscr{S}$ if and only if, for every pair of t-subsets of $\mathscr{S}$ (s_1, s_2, say), there exist precisely w permutations e in $\mathscr{E}$ such that $e(s_1) = s_2$.*

By Theorem 9.3, the study of a $U(t)$-secrecy scheme having

$$b = \binom{k}{t}$$

is then precisely equivalent to the study of $\{t,1\}$-homogeneous sets of permutations on a set of size k. Moreover, by Theorem 9.4, the study of a $S(t)$-secrecy scheme having

$$b = \binom{k}{t}$$

is precisely equivalent to the study of $\{t,1\}$-homogeneous sets of permutations on a set of size k which is $\{r, \binom{k}{t}/\binom{k}{r}\}$-homogeneous for every r $(1 \leqslant r \leqslant t)$ as well.

LEMMA 9.7

If $\mathscr{E}$ is $\{t,w\}$-homogeneous on $\mathscr{S}$, then $\mathscr{E}$ is also $\{k-t, w\}$-homogeneous on $\mathscr{S}$ where $|\mathscr{S}| = k$.

PROOF Suppose $\mathscr{E}$ is $\{t, w\}$-homogeneous on $\mathscr{S}$, and $s_{k-t}^{(1)}$, $s_{k-t}^{(2)}$ are $(k-t)$-subsets of $\mathscr{S}$. Then it should be clear that if

$$s_t^{(i)} = \mathscr{S} \setminus s_{k-t}^{(i)} \quad (i = 1, 2),$$

then encoding rule e satisfies

$$e(s_{k-t}^{(1)}) = e(s_{k-t}^{(2)})$$

if and only if

$$e(s_t^{(1)}) = s_t^{(2)}.$$

The result follows. □

LEMMA 9.8

If $\mathscr{E}$ is $\{t,w\}$-homogeneous, then

$$b = w \cdot \binom{k}{t}.$$

PROOF Let $s_t^{(1)}$ be a fixed t-subset of $\mathscr{S}$. Then for any t-subset of $\mathscr{S}$ ($s_t^{(2)}$, say) there exist precisely w permutations in $\mathscr{E}$ mapping $s_t^{(1)}$ onto $s_t^{(2)}$. Since there are $\binom{k}{t}$ such subsets $s_t^{(2)}$, and since each element of $\mathscr{E}$ must map $s_t^{(1)}$ onto one of such t-subsets, the result follows. □

LEMMA 9.9 (Mowbray)

If $\mathscr{E}$ is $\{t,w\}$-homogeneous on $\mathscr{S}$, where $1 \leqslant t \leqslant (k+1)/2$ $(|\mathscr{S}| = k)$, then $\mathscr{S}$ is also $\{t', w'\}$-homogeneous on $\mathscr{S}$ for every $t' \leqslant t$, where

$$w' = w \cdot \binom{k}{t} \Big/ \binom{k}{t'}.$$

198

Authentication Codes and Combinatorial Designs

PROOF Suppose that $\mathscr{E}$ is $\{t, w\}$-homogeneous on $\mathscr{S}$. If X, Y are any pair of $(t-1)$-subsets of $\mathscr{S}$ and $0 \leqslant s \leqslant t-1$, then let $N(X, Y, s)$ denote the number of permutations $e \in \mathscr{E}$ such that $e(X)$ and Y have precisely s elements in common. We now show (by induction on s) that, for all $0 \leqslant s \leqslant t-1$,

$$N(X, Y, s) = \frac{t-s}{t}\binom{t-1}{s} \cdot \frac{k-(t-1)}{k-2(t-1)+s}\binom{k-(t-1)}{t-s}w. \qquad (9.3)$$

First, suppose that $s = 0$. If e is such that $e(X)$ and Y are disjoint, then there exist precisely $k - 2(t-1) = |\mathscr{S} \setminus \{X \cup e^{-1}(Y)\}|$ choices for a t-set X' which contains X and $e(X')$ is disjoint from Y. That is, there are exactly $(k-2[t-1]) \cdot N(X, Y, 0)$ pairs (e, X'), where $|X'| = t$, X' contains X, and $e(X')$ and Y are disjoint. On other hand, there exist precisely $k - (t-1) = |\mathscr{S} \setminus X|$ choices for a t-set X' which contains X, and precisely $\binom{k-[t-1]}{t}$ choices for a t-set $Y' \subset \mathscr{S} \setminus Y$ which is disjoint from Y. For a pair of X' and Y' chosen above, there are w encoding rules mapping X' onto Y'. Hence, there are $(k - [t-1])\binom{k-[t-1]}{t}w$ such pairs (e, X'). Therefore,

$$N(X, Y, 0) = \frac{k-(t-1)}{k-2(t-1)}\binom{k-(t-1)}{t}w$$

(where $k - 2[t-1] > 0$, since $t \leqslant [k+1]/2$) and Equation (9.3) is true for $s = 0$.

Now suppose that Equation (9.3) is true for $s = r-1$. First, suppose that e is such that $e(X)$ and Y have precisely r elements in common; then there exist precisely $(k - 2[t-1] + r) = |\mathscr{S} \setminus \{X \cup e^{-1}(Y)\}|$ choices for a t-set X' which contains X and $e(X')$ and Y meet in precisely r elements. Second, suppose that e is such that $e(X)$ and Y have precisely $r-1$ elements in common; then there exist precisely $t - r = |e^{-1}(Y) \setminus \{e^{-1}(Y) \cap X\}|$ choices for a t-set X' which contains X and $e(X')$ and Y meet in precisely r elements. That is, there are exactly

$$(k - 2(t-1) + r) \cdot N(X, Y, r) + (t-r) \cdot N(X, Y, r-1)$$

pairs (e, X'), where $|X'| = t$, X' contains X, and $e(X')$ and Y meet in precisely r elements of $\mathscr{S}$. On other hand, there exist $k - (t-1) = |\mathscr{S} \setminus X|$ choices for a t-set X' which contains X, and precisely $\binom{k-[t-1]}{t-r}\binom{t-1}{r}$ choices for a t-set Y' which meets Y in exactly r elements. Since $\mathscr{E}$ is t-homogeneous, there are $(k - [t-1])\binom{k-[t-1]}{t-r}\binom{t-1}{r}w$ such pairs (e, X'). Therefore,

$$(k - 2(t-1) + r) \cdot N(X, Y, r) = (k - (t-1))\binom{k-(t-1)}{t-r}\binom{t-1}{r}w$$
$$- (t-r)N(X, Y, r-1).$$

By induction hypothesis, we have

$$(k-2(t-1)+r) \cdot N(X,Y,r) = w(k-(t-1)) \left[\binom{t-1}{r} \binom{k-(t-1)}{t-r} \right.$$

$$\left. - \frac{(t-r+1)(t-r)}{t} \binom{t-1}{r-1} \binom{k-(t-1)}{t-r+1} \middle/ (k-2(t-1)+r-1) \right]$$

$$= w(k-(t-1)) \left[\frac{(t-1)!(k-(t-1))!}{r!(t-r-1)!(t-r)!(k-2(t-1)+r-1)!} \right.$$

$$\left. - \frac{(t-r+1)(t-r)(t-1)!(k-(t-1))!}{t(r-1)!(t-r)!(t-r+1)!(k-2(t-1)+r-2)!(k-2(t-1)+r-1)} \right]$$

$$= w(k-(t-1)) \binom{t-1}{r} \binom{k-(t-1)}{t-r} \left(1 - \frac{r}{t} \right).$$

Thus Equation (9.3) is true for $s = r$, it is also true for all $0 \leqslant s \leqslant t-1$ by induction. Setting $s = t-1$ in Equation (9.3) we have

$$N(X,Y,t-1) = w \cdot \frac{k-(t-1)}{t} = w \cdot \binom{k}{t} \middle/ \binom{k}{t-1}.$$

Since X and Y are any pair of $(t-1)$-subsets of $\mathcal{S}$, it follows that $\mathcal{E}$ is $(t-1, w')$-homogeneous on $\mathcal{S}$. The result follows. □

The result of Lemma 9.9, then taken in conjunction with Theorem 9.4 and Lemma 9.7, implies the following.

COROLLARY 9.1
If $\mathcal{A} = (\mathcal{S}, \mathcal{M}, \mathcal{E}, p_S, p_E)$, with $|\mathcal{M}| = |\mathcal{S}| = k$, $|\mathcal{E}| = \binom{k}{t}$, provides $U(t)$-secrecy and $1 \leqslant t \leqslant (k+1)/2$, then it also provides $U(r)$-secrecy for every r satisfying either $1 \leqslant r \leqslant t$ or $k-t \leqslant r \leqslant k$.

The problem remains of constructing $\{t,1\}$-homogeneous sets of permutations by Theorem 9.1. We first note the following result, giving a necessary condition for the existence of an $\{t,1\}$-homogeneous set.

LEMMA 9.10
If $\mathcal{E}$ is $\{t,1\}$-homogeneous on the k-set $\mathcal{S}$ $\{1 \leqslant t \leqslant (k+1)/2\}$, then $\binom{k}{t'}$ is a factor of $\binom{k}{t}$ for every t' $(1 \leqslant t' \leqslant t)$.

PROOF This lemma follows immediately from Lemma 9.9. □

We now consider construction of $U(t)$-secrecy systems which satisfy

$$|\mathcal{E}| = \binom{k}{t}.$$

For the case $t = 1$, as observed by Shannon [40], the existence of such a set is precisely equivalent to the existence of a Latin square of order k. A Latin square of order k is merely a $k \times k$ matrix, all of whose entries are taken from the set $\{1, 2, \cdots, k\}$ with the property that the entries in any row are all distinct and the entries in any column are all distinct. Each row (and each column) will, therefore, contain a permutation of the numbers 1 to k. If the row i contains the entries $r_1, r_2, \cdots, r_k$, then define the permutation ρ_i by

$$\rho_i(j) = r_j.$$

It is then clear that the k permutations $\rho_1, \rho_2, \cdots, \rho_k$ will form a $\{1, 1\}$-homogeneous set on $\{1, 2, \cdots, k\}$. Moreover, any $\{1, 1\}$-homogeneous set can be used to derive a Latin square. It should also be clear that the one-time pad cipher is equivalent to a Latin square. It is easy to construct a Latin square of any desired size (e.g., by letting the first row be any permutation and letting the subsequent rows be defined as all cyclic shifts of the first row), and hence $\{1, 1\}$-homogeneous sets exist for all values of k.

Finally, note that, since $U(1)$-secrecy, $S(1)$-secrecy, and $O(1)$-secrecy are all equivalent, Latin squares also precisely correspond to perfect schemes of other types of perfect secrecy.

For the case $t = 2$, we assert that the existence of a $\{2, 1\}$-homogeneous set of permutations is equivalent to the existence of a Perpendicular Array (**PA**) with parameters $PA(k, k)$.

DEFINITION 9.4   *A Perpendicular Array of order k and depth s {written $PA(k, s)$} is a $\binom{k}{2} \times s$ array $X = (x_{ij})$ with entries from a set $\mathcal{M}$ of k elements such that, for any two columns of X, the $\binom{k}{2}$ rows contain all $\binom{k}{2}$ unordered pairs of distinct elements of $\mathcal{M}$.*

If $s = k$, i.e., when we have a $PA(k, k)$, then X is a $\binom{k}{2} \times k$ array, and it is straightforward to see that each row of X is a permutation of the elements of $\mathcal{M}$. We thereby derive a set of $\binom{k}{2}$ permutations which forms a $\{2, 1\}$-homogeneous set on $\mathcal{M}$. Conversely, given any $\{2, 1\}$-homogeneous set of permutations on a set of size k, we may immediately derive a $PA(k, k)$.

If $k \geqslant 3$ and k is odd, any $\{2, 1\}$-homogeneous set of permutations on a set of size k is also a $\{1, [k - 1]/2\}$-homogeneous set by Lemma 9.9. Hence, by Theorem 9.4, the existence of an $S(2)$-secrecy system with

$$b = \binom{k}{2}$$

is also equivalent to the existence of a $PA(k, k)$.

It is well known [25] that if

$$k = p^a \qquad (a \geqslant 1, \ p \text{ an odd prime})$$

then there exists a $PA(k, k)$.

We now relax our requirement that $v = k$, and consider schemes for which

$$b = (v/k) \cdot \binom{k}{t}, \quad \text{and} \quad v \geqslant k.$$

We first consider the special case $t = 1$, thus the above equality reduces to $b = v$. Using Lemma 9.4 it is then straightforward to see that the existence of such a set of encoding rules is precisely equivalent to the existence of a $k \times v$ Latin rectangle, where a Latin rectangle is merely a $k \times v$ matrix, all of whose entries are taken from the set $\{1, 2, \cdots, v\}$ with the property that the entries in any row or column are all distinct (hence, $k \leqslant v$). It is easy to construct a Latin rectangle of any k and v, the same as in the case of $v = k$.

Note that when $k < v$ the scheme constructed by Latin rectangle, taking each column as an encoding rule, is the counterexample of Theorem 9.1 (i) and Theorem 9.2 (i) for the case of $t = 1$.

If $t > 1$, then, by Theorem 9.1 (i), $v = kr$ for some integer r, and the message set $\mathcal{M}$ can be partitioned into r subsets of size of k, say $\mathcal{M}_1, \mathcal{M}_2, \cdots, \mathcal{M}_r$, such that, for any encoding rule e,

$$e(\mathcal{S}) = \mathcal{M}_i$$

for some i. Therefore, let $\mathcal{E}_i$ denote the set of encoding rules mapping $\mathcal{S}$ onto $\mathcal{M}_i$ and then $\mathcal{E}_1, \mathcal{E}_2, \cdots, \mathcal{E}_r$ will form a partition of $\mathcal{E}$. Moreover, from Theorem 9.1 (ii), if

$$e_1(\mathcal{S}) = e_2(\mathcal{S}),$$

then

$$p(e_1) = p(e_2),$$

and hence, let p_i denote the probability $p(e)$ for any e in $\mathcal{E}_i$. It is then straightforward to establish that each triple $(\mathcal{S}, \mathcal{M}_i, \mathcal{E}_i)$ forms an $(t, 1)$-homogeneous set of permutations. This means that the study of $U(t)$-secrecy schemes with

$$b = (v/k) \cdot \binom{k}{t}, \quad \text{and} \quad v > k$$

is contained within the study of such schemes with $v = k$.

We now study the perfect systems providing $O(t)$-secrecy. First, we examine the case where $v = k$ as above, then $b = k!/(k - t)!$ by Theorem 9.2. We may identify $\mathcal{S}$ with $\mathcal{M}$, and each encoding rule is then no more than a permutation on $\mathcal{M}$. Moreover, by Theorem 9.2 (ii), each encoding rule must be equiprobable.

THEOREM 9.5

The scheme $\mathcal{A} = (\mathcal{S}, \mathcal{M}, \mathcal{E}, p_S, p_E)$ with $|\mathcal{S}| = |\mathcal{M}| = k$, $b = k!/(k - t)!$ provides $O(t)$-secrecy if and only if

(i) $p(e) = (k-t)!/k!$ *for every encoding rule* e.

(ii) *For every pair of t-tuples of distinct elements* m^t, s^t *of* $\mathcal{M}$, *there exists a unique encoding rule* e *with*

$$e(s^t) = m^t.$$

PROOF   Suppose that the scheme provides $O(t)$-secrecy. Then (i) holds by Theorem 9.2 (ii). Moreover, (ii) holds by Lemma 9.6 (i). Now suppose (i) and (ii) hold and suppose s^t, m^t are two t-tuples of $\mathcal{M}$. Then

$$
\begin{aligned}
p(s^t|m^t) &= p(m^t|s^t)p(s^t)/p(m^t) \\
&= p(e)p(s^t)/p(m^t) &&\{\text{by } (ii)\} \\
&= p(s^t)(k-t)!/(k!p(m^t)), &&\{\text{by } (i)\}
\end{aligned}
$$

But, by definition

$$p(m^t) = \sum_{s^t \in \mathscr{S}^t} p(s^t) \sum_{e:\, e(s^t)=m^t} p(e) = \sum_{s^t \in \mathscr{S}^t} p(s^t) \cdot (k-t)!/k! = (k-t)!/k!.$$

Hence, as required

$$p(s^t|m^t) = p(s^t).$$

and the result follows.   □

Note that, for the case $t = 1$, the above theorem coincides with Theorem 9.3.

THEOREM 9.6
Suppose $\mathcal{E}$ is a set of encoding rules (permutations) for a scheme $\mathcal{A}$ with $|\mathscr{S}| = |\mathcal{M}| = k$ and $b = k!/(k-t)!$ which provides $O(t)$-secrecy. Then, for every pair of r-tuples ($r \leqslant t$) of distinct elements s^r, m^r, there exist precisely b' encoding rules e with

$$e(s^r) = m^r,$$

where

$$b' = (k-r)!/(k-t)!.$$

PROOF   First, we observe that, by Lemma 9.1, any scheme providing $O(t)$-secrecy must also provide $O(r)$-secrecy for every $r \leqslant t$. Suppose s^r and m^r are any r-tuples of distinct elements of $\mathcal{M}$ and let $\mathcal{E}(s^r, m^r)$ denote the set of encoding rules which map s^r onto m^r. By definition, since the scheme provides $O(r)$-secrecy,

$$p(s^r | m^r) = p(s^r).$$

Hence,

$$p(m^r) = p(m^r | s^r) = \sum_{e \in \mathscr{E}(s^r, m^r)} p(e) = |\mathscr{E}(s^r, m^r)| \cdot (k - t)! / k!. \qquad (9.4)$$

Now, if we fix m^r and let s^r range over all $k!/(k-r)!$ possible r-tuples of distinct elements of $\mathscr{S}$, then the sets $\mathscr{E}(s^r, m^r)$ will be pairwise disjoint and their union is $\mathscr{E}$. Moreover, since the left-hand side of Equation (9.4) will be fixed, then all must have the same size. Hence,

$$|\mathscr{E}(s^r, m^r)| = |\mathscr{E}|(k - r)!/k! = (k - r)!/(k - t)!.$$

as required. □

DEFINITION 9.5 *Suppose $\mathscr{E}$ is a set of permutations on the set $\mathscr{S}$. Then $\mathscr{E}$ is said to be $\{t, w\}$-transitive on $\mathscr{S}$ if and only if, for every pair of t-tuples of distinct elements of $\mathscr{S}$ (s_1, s_2 say), there exist precisely w permutations e in $\mathscr{E}$ such that*

$$e(s_1) = s_2.$$

By Theorem 9.5, the study of $O(t)$-secrecy scheme having $b = k!/(k-t)!$ is then precisely equivalent to the study of $\{t, 1\}$-transitive sets of permutations on a set of size k. Moreover, by Theorem 9.6, every $\{t, 1\}$-transitive set of permutations is also $\{r, (k - r)!/(k - t)!\}$-transitive for every r satisfying $1 \leqslant r \leqslant t$.

The problem remains of constructing $\{t, 1\}$-transitive sets of permutations. The case $t = 1$ has already been studied above.

For the case $t \geqslant 2$, the theory of finite groups provides a number of examples. Suppose $\mathscr{E}$ is a set of permutations on k elements. If $\mathscr{E}$ forms a subgroup of the symmetric group S_k, then $\mathscr{E}$ is a $\{t, w\}$-transitive set of permutations if and only if it is a t-transitive group. Moreover, using the language of group theory, it is a $\{t, 1\}$-transitive set if and only if it is a sharply t-transitive group (see, for example, Reference [56]).

Two "trivial" families of group-based examples are provided by the symmetric group S_k and the alternating group A_k. It is known [22] that S_k is sharply k-transitive and A_k is sharply $(k - 2)$-transitive, hence they provide two examples of perfect systems providing $O(k)$ and $O(k-2)$-secrecy, respectively.

Sharply 2- and 3-transitive groups are known to exist for infinitely many values of k. However, the situation is very different for $t \geqslant 4$. Apart from S_k and A_k, the only t-transitive groups with $t \geqslant 4$ are the Mathieu groups: M_{11}, M_{12}, M_{23}, and M_{24}, where M_i acts on a set of i elements [24]. Groups M_{11} and M_{23} are 4-transitive and M_{12} and M_{24} are 5-transitive; M_{11} and

M_{12} are sharply 4- and 5-transitive, respectively, whereas M_{23} and M_{24} are not. Hence, M_{11} and M_{12} (of order 7920 and 95040, respectively) are the only "nontrivial" examples of sharply t-transitive groups for $t \geqslant 2$.

To obtain further examples of $\{t, 1\}$-transitive sets, it is, therefore, necessary to look for examples where the set of permutations does not form a group. There are many examples of $\{t, 1\}$-transitive sets which are not subgroups of S_k [2].

To conclude this discussion of $O(t)$-secrecy systems, we now relax our requirement that $v = k$, and consider schemes for which

$$b = (v/k) \cdot k!/(k - t)! \qquad \text{and} \qquad v \geqslant k.$$

The case $t = 1$ coincides with the discussion of $U(1)$-secrecy above.

If $t > 1$, then, by Theorem 9.2 (i), $v = kr$ for some integer r, and the message set $\mathcal{M}$ can be partitioned into r subsets of size k, say $\mathcal{M}_1, \mathcal{M}_2, \cdots, \mathcal{M}_r$, such that, for any encoding rule e,

$$e(\mathcal{S}) = \mathcal{M}_i$$

for some i. Therefore, let $\mathcal{E}_i$ denote the set of encoding rules mapping $\mathcal{S}$ onto $\mathcal{M}_i$ and then $\mathcal{E}_1, \mathcal{E}_2, \cdots, \mathcal{E}_r$ will form a partition of $\mathcal{E}$. Moreover, by Theorem 9.2 (ii), if

$$e_1(\mathcal{S}) = e_2(\mathcal{S}),$$

then

$$p(e_1) = p(e_2),$$

and hence, let p_i denote the probability $p(e)$ for any e in $\mathcal{E}_i$.

It is then straightforward to establish that each triple $(\mathcal{S}, \mathcal{M}_i, \mathcal{E}_i)$ forms a $\{t, 1\}$-transitive set of permutations. This means that the study of $O(t)$-secrecy schemes with

$$b = (v/k) \cdot k!/(k - t)! \quad \text{and} \quad v > k$$

is contained within the study of such schemes with $v = k$.

9.3   Authentication Schemes with Perfect Secrecy

We studied the system which provides only secrecy in the above two sections. Now we study the systems that offer both authentication and secrecy in the next two sections. More precisely, we study the systems which are key-entropy minimal of order t and provide $U(t)$-secrecy or $O(t)$-secrecy as well. The most important subject of concern is the bounds of encoding rules for these systems. Clearly, it will satisfy both the bounds given in Equation 3.9 of Chapter 3 and those given in Theorem 9.1 or Theorem 9.2.

Let p_{S^t} and p_{S_t} denote the probability distributions of the random variables S^t and S_t, respectively. For a t-tuple $s^t = (s_1, s_2, \cdots, s_t)$, we denote its corresponding subset $\{s_1, s_2, \cdots, s_t\}$ by $[s^t]$.

DEFINITION 9.6 *The probability distribution of p_{S^t} is said to be set-uniform if it is uniform on $\{u^t \in \mathscr{S}^t \,|\, [u^t] = s_t'\}$ for each $s_t' \in \mathscr{S}_t$.*

LEMMA 9.11
Suppose that p_{S^t} is set-uniform and $m^t \in \overline{\mathscr{M}^t}$, $e \in \mathscr{E}(m^t)$, then $p(m^t|e)$ and $p(m^t)$ are constant on $\{m^t \in \overline{\mathscr{M}^t} \mid [m^t] = m_t'\}$ for any $m_t' \in \overline{\mathscr{M}_t}$.

PROOF Assume that $m_1^t, m_2^t \in \overline{\mathscr{M}^t}$ and $[m_1^t] = [m_2^t]$, then $\mathscr{E}(m_1^t) = \mathscr{E}(m_2^t)$ and $[f_e(m_1^t)] = [f_e(m_2^t)]$ for any $e \in \mathscr{E}(m_1^t)$. By the assumption of the lemma,

$$p(m_1^t|e) = p(f_e(m_1^t)) = p(f_e(m_2^t)) = p(m_2^t|e).$$

Furthermore,

$$p(m_1^t) = \sum_{e \in \mathscr{E}(m_1^t)} p(e)p(f_e(m_1^t))$$
$$= \sum_{e \in \mathscr{E}(m_2^t)} p(e)p(f_e(m_2^t)) = p(m_2^t).$$

Thus, the result follows. ∎

LEMMA 9.12
Suppose p_{S^t} is set-uniform, then

$$H(E|M_t) = H(E|M^t)$$

where M_t and M^t denote the random variables defined on the sets $\overline{\mathscr{M}_t}$ and $\overline{\mathscr{M}^t}$, respectively.

PROOF Assume that $m^t \in \overline{\mathscr{M}^t}$, $[m^t] = m_t' \in \overline{\mathscr{M}_t}$ and $e \in \mathscr{E}(m^t)$. By Lemma 9.11 we have

$$p(m_t') = \sum_{u^t \in \overline{\mathscr{M}^t}:\, [n^t]=m_t'} p(u^t) = t!p(m^t),$$
$$p(m_t'|e) = \sum_{u^t \in \overline{\mathscr{M}^t}:\, [n^t]=m_t'} p(u^t|e) = t!p(m^t|e)$$
$$p(e, m_t') = p(e)p(m_t'|e) = t!p(e, m^t).$$

and

$$p(e|m^t) = \frac{p(e, m^t)}{p(m^t)} = \frac{p(e, m_t')}{p(m_t')} = p(e|m_t')).$$

Hence,

$$
\begin{aligned}
H(E|M^t) &= - \sum_{e \in \mathscr{E}, m^t \in \overline{\mathscr{M}}^t} p(e, m^t) \log p(e|m^t) \\
&= - \sum_{e \in \mathscr{E}, m_t' \in \overline{\mathscr{M}}_t} \sum_{m^t : [m^t] = m_t'} p(e, m^t) \log p(e|m^t) \\
&= - \sum_{e \in \mathscr{E}, m_t' \in \overline{\mathscr{M}}_t} p(e, m_t') \log p(e|m_t') \\
&= H(E|M_t)
\end{aligned}
$$

and the result follows.   ⬜

LEMMA 9.13
Suppose $\mathscr{A} = (\mathscr{S}, \mathscr{M}, \mathscr{E}, p_S, p_E)$ is t-fold key-entropy minimal ($t \leqslant k$). Then, for any $m^r \in \mathscr{M}^r, (1 \leqslant r \leqslant t)$, $p(m^r|e)$ is constant for all $e \in \mathscr{E}(m^r)$.

PROOF   This lemma is nothing but the (i) of Corollary 3.2, i.e., p_{S^r} is message uniform.   ⬜

PROPOSITION 9.1
Suppose $\mathscr{A} = (\mathscr{S}, \mathscr{M}, \mathscr{E}, p_S, p_E)$ is t-fold key-entropy minimal ($t \leqslant k$).

(i) *If $\mathscr{A}$ has $O(t)$-secrecy then p_{S^t} is uniform.*

(ii) *If $\mathscr{A}$ has $U(t)$-secrecy and p_{S^t} is set-uniform, then p_{S_t} is uniform (and p_{S^t} is uniform too).*

PROOF   (i) Let m^t be an allowable t-tuple of distinct messages and s_1^t, s_2^t be any pair of t-tuples of distinct source states. As $\mathscr{A}$ provides $O(t)$-secrecy, there exist encoding rules e_1 and e_2 such that $e_1(s_1^t) = m^t$, $e_2(s_2^t) = m^t$ by Lemma 9.5. Using Lemma 9.13, $p(s_1^t) = p(m^t|e_1) = p(m^t|e_2) = p(s_2^t)$. The result follows.

(ii) Let m_t' be an allowable t-subset of distinct messages and s_t', u_t' be any pair of t-subsets of distinct source states. As $\mathscr{A}$ provides $U(t)$-secrecy, there exist encoding rules e_1 and e_2 such that $e_1(s_t') = m_t'$ and $e_2(u_t') = m_t'$ by Lemma 9.3. Let s^t be a t-tuple of distinct source states such that $[s^t] = s_t'$, then $e_1(s^t) = m^t$ for some t-tuple m^t with $[m^t] = m_t'$. There also exists a t-tuple u^t such that $e_2(u^t) = m^t$ and $[u^t] = u_t'$. Since p_{S^t} is set uniform, we have

$$p(s_t') = (t!)p(s^t) = (t!)p(m^t|e_1) = (t!)p(m^t|e_2) = (t!)p(u^t) = p(u_t').$$

Hence, P_{S_t} is uniform.

Furthermore, let s_1^t, $s_2^t \in \mathscr{S}^t$, then

$$p(s_1^t) = \frac{1}{t!}p([s_1^t]) = \frac{1}{t!}p([s_2^t]) = p(s_2^t),$$

hence, p_{S^t} is uniform too. ▯

LEMMA 9.14
If $\mathscr{A} = (\mathscr{S},\mathscr{M},\mathscr{E},p_S,p_E)$ has $O(t)$-secrecy, then

$$H(S^t|M^t) = H(S^t).$$

If $\mathscr{A}$ has $U(t)$-secrecy then

$$H(S_t|M_t) = H(S_t).$$

PROOF Suppose that $\mathscr{A}$ has $O(t)$-secrecy. By Definition 9.2

$$H(S^t|M^t) = -\sum_{s^t,m^t} p(s^t,m^t) \log p(s^t|m^t)$$

$$= -\sum_{s^t,m^t} p(s^t,m^t) \log p(s^t)$$

$$= -\sum_{s^t} p(s^t) \log p(s^t) = H(S^t).$$

Similarly, the equality in the lemma for $U(t)$-secrecy can be proved. ▯

THEOREM 9.7
Let $\mathscr{A} = (\mathscr{S},\mathscr{M},\mathscr{E},p_S,p_E)$ and $1 \leqslant t \leqslant k$. Suppose that $\mathscr{A}$ provides $O(t)$-secrecy, then

$$|\mathscr{E}| \geqslant \frac{k!}{(k-t)!} \prod_{r=0}^{t-1} P_r^{-1}.$$

The equality holds if and only if $\mathscr{A}$ is t-fold key-entropy minimal, p_E and p_{S^t} are uniform, and $H(E|S^t,M^t) = 0$ {which means for each $m^t \in \mathscr{M}^t$ and each $s^t \in \mathscr{S}^t$, there is a unique $e \in \mathscr{E}$ such that $e(s^t) = m^t$}.

PROOF We have

$$P_r \geqslant 2^{H(E|M^{r+1})-H(E|M^r)}, \qquad 0 \leqslant r \leqslant t-1$$

by Proposition 3.1. Hence,

$$\prod_{r=0}^{t-1} P_r \geqslant 2^{H(E|M^t)-H(E)}, \tag{9.5}$$

The equality holds if and only if $\mathscr{A}$ is t-fold key-entropy minimal. It follows that

$$2^{H(E)} \geqslant 2^{H(E|M^t)} \prod_{r=0}^{t-1} P_r^{-1}. \tag{9.6}$$

By Lemma 3.2,

$$H(E, S^t|M^t) = H(E|M^t) + H(S^t|E, M^t) = H(S^t|M^t) + H(E|S^t, M^t).$$

Since $H(S^t|E, M^t) = 0$ and $\mathscr{A}$ provides $O(t)$-secrecy,

$$H(E|M^t) = H(S^t|M^t) + H(E|S^t, M^t) \geqslant H(S^t|M^t) = H(S^t). \tag{9.7}$$

Here, Lemma 9.14 is used in the last equality. The inequality in Equation (9.7) becomes equality if and only if $H(E|S^t, M^t) = 0$. Therefore,

$$|\mathscr{E}| \geqslant 2^{H(E)} \geqslant 2^{H(S^t)} \prod_{r=0}^{t-1} P_r^{-1} \tag{9.8}$$

by Equations (9.6) and (9.7).

If the equality in Equation (9.8) holds, then p_E is uniform and the equalities in Equations (9.5) and (9.7) hold, i.e., $\mathscr{A}$ is t-fold key-entropy minimal and $H(E|S^t, M^t) = 0$. By Proposition 9.1 (i), p_{S^t} is uniform, hence,

$$2^{H(S^t)} = |\mathscr{S}^t| = k!/(k-t)! \tag{9.9}$$

and

$$|\mathscr{E}| = \frac{k!}{(k-t)!} \prod_{r=0}^{t-1} P_r^{-1}. \tag{9.10}$$

Now suppose that $\mathscr{A}$ is t-fold key-entropy minimal, p_E and p_{S^t} are uniform, and $H(E|S^t, M^t) = 0$. By Equations (9.5), (9.7), (9.8), and (9.9), the equality in Equation (9.10) holds. The theorem is proved. □

THEOREM 9.8

Let $\mathscr{A} = (\mathscr{S}, \mathscr{M}, \mathscr{E}, p_S, p_E)$ and $1 \leqslant t \leqslant k$. Suppose that $\mathscr{A}$ provides $U(t)$-secrecy and p_{S^t} is set-uniform. Then

$$|\mathscr{E}| \geqslant \binom{k}{t} \prod_{r=0}^{t-1} P_r^{-1}.$$

The equality holds if and only if $\mathscr{A}$ is t-fold key-entropy minimal, p_E and $p_{S'}$ are uniform, and $H(E|S_t, M_t) = 0$ (which means that for each $m'_t \in \overline{\mathscr{M}}_t$ and each $s'_t \in \mathscr{S}_t$ there is a unique $e \in \mathscr{E}$ such that $e(s'_t) = m'_t$).

PROOF Since $p_{S'}$ is set-uniform, by Lemma 9.12 and Equation (9.6),

$$2^{H(E)} \geqslant 2^{H(E|M_t)} \prod_{r=0}^{t-1} P_r^{-1}. \tag{9.11}$$

The equality holds if and only if $\mathscr{A}$ is t-fold key-entropy minimal. Similarly, we have

$$H(E, S_t|M_t) = H(E|M_t) + H(S_t|E, M_t) = H(S_t|M_t) + H(E|S_t, M_t).$$

Since $H(S_t|E, M_t) = 0$ and $\mathscr{A}$ has $U(t)$-secrecy, by Lemma 9.14

$$H(E|M_t) = H(S_t|M_t) + H(E|S_t, M_t) \geqslant H(S_t|M_t) = H(S_t). \tag{9.12}$$

The inequality in Equation (9.12) becomes equality if and only if $H(E|S_t, M_t) = 0$. Therefore,

$$|\mathscr{E}| \geqslant 2^{H(E)} \geqslant 2^{H(S_t)} \prod_{r=0}^{t-1} P_r^{-1}. \tag{9.13}$$

If the equality in Equation (9.13) holds, then p_E is uniform and the equalities in Equations (9.11) and (9.12) hold, i.e., $\mathscr{A}$ is t-fold key-entropy minimal and $H(E|S_t, M_t) = 0$. By Proposition 9.1 (ii), P_{S_t} is uniform (hence, $p_{S'}$ is uniform too).

$$2^{H(S_t)} = |\mathscr{S}_t| = \binom{k}{t} \tag{9.14}$$

and

$$|\mathscr{E}| = \binom{k}{t} \prod_{r=0}^{t-1} P_r^{-1} \tag{9.15}$$

Now suppose that $\mathscr{A}$ is t-fold key-entropy minimal, p_E and $p_{S'}$ are uniform, and $H(E|S_t, M_t) = 0$. Obviously P_{s_t} is uniform when $p_{S'}$ is uniform. By Equations (9.11), (9.12), (9.13), and (9.14), the equality in Equation (9.15) holds. Thus, the theorem is proved. □

9.4 Construction of Perfect Authentication/Secrecy Schemes

We now establish a combinatorial characterization similar to Theorem 3.1 for authentication/secrecy schemes meeting the bound of Theorem 9.7 or Theorem 9.8.

THEOREM 9.9
Suppose that the scheme $\mathscr{A} = (\mathscr{S}, \mathscr{M}, \mathscr{E}, p_S, p_E)$ has $O(t)$-secrecy. Then the number of encoding rules satisfies

$$|\mathscr{E}| = k!/(k-t)! \prod_{r=0}^{t-1} P_r \tag{9.16}$$

if and only if the pair

$$(\mathscr{M}, \{\mathscr{M}(e) \mid e \in \mathscr{E}\}) \tag{9.17}$$

is a strong partially balanced design (SPBD) t-$(v, b, k; \lambda_1, \lambda_2, \cdots, \lambda_t, 0)$ with $\lambda_t = k!/(k-t)!$, and p_E and $p_{S'}$ are uniform distributions. Here,

$$v = |\mathscr{M}|, b = |\mathscr{E}|, k = |\mathscr{S}|, \lambda_r = \frac{k!}{(k-t)!}(P_r P_{r+1} \cdots P_{t-1})^{-1} \quad (1 \leqslant r \leqslant t-1).$$

PROOF Suppose that $\mathscr{A}$ has $O(t)$-secrecy and $|\mathscr{E}|$ satisfies Equation (9.16). By Theorem 9.7 p_E and $p_{S'}$ are uniform, and $H(E|S^t, M^t) = 0$, which means that for each $m^t \in \mathscr{M}^t$ and each $s^t \in \mathscr{S}^t$, there is a unique $e \in \mathscr{E}$ such that $e(s^t) = m^t$. Hence, for each $m^t \in \mathscr{M}^t$, $\lambda_t = |\mathscr{E}(m^t)| = |\mathscr{S}^t| = k!/(k-t)!$.

Since $\mathscr{A}$ is t-fold key-entropy minimal and p_E is uniform, we have, by Corollary 3.2 (ii),

$$\lambda_r = |\mathscr{E}(m^r)| = |\mathscr{E}| \cdot P_0 P_1 \cdots P_{r-1} = \frac{k!}{(k-t)!}(P_r P_{r+1} \cdots P_{t-1})^{-1}$$

where m^r is any element of $\overline{\mathscr{M}^r}$ $(0 \leqslant r \leqslant t-1)$. Hence, the pair in Equation (9.17) is a SPBD t-$(v, b, k; \lambda_1, \lambda_2, \cdots, \lambda_t, 0)$ with the required parameters.

Conversely, we assume that the pair Equation (9.17) is a SPBD t-$(v, b, k; \lambda_1, \lambda_2, \cdots, \lambda_t, 0)$ with the given parameters and p_E, $p_{S'}$ are uniform. For any $m^r * m \in \overline{\mathscr{M}^{r+1}}$ $(0 \leqslant r \leqslant t-1)$ and any $e \in \mathscr{E}(m^r * m)$ we have, for $0 \leqslant r \leqslant t-1$,

$$\frac{p(e|m^r)}{p(e|m^r * m)} = \frac{p(e, m^r)p(m^r * m)}{p(e, m^r * m)p(m^r)}$$

$$= \frac{p(f_e(m^r)) \displaystyle\sum_{e' \in \mathscr{E}(m^r * m)} p(e')p(f_{e'}(m^r * m))}{p(f_e(m^r * m)) \displaystyle\sum_{e' \in \mathscr{E}(m^r)} p(e')p(f_{e'}(m^r))}$$

$$= \frac{|\mathscr{E}(m^r * m)|}{|\mathscr{E}(m^r)|} = \frac{\lambda_{r+1}}{\lambda_r}$$

where $\lambda_0 = |\mathscr{E}| = b$. The above ratios do not depend on $m^r * m$ and e, hence $P_r (0 \leqslant r \leqslant t-1)$ achieve their information-theoretic bounds (Proposition 3.1) and

$$P_r = \frac{\lambda_{r+1}}{\lambda_r} \qquad 0 \leqslant r \leqslant t-1.$$

Hence,

$$b = \lambda_0 = \lambda_1/P_0 = \frac{k!}{(k-t)!}(P_0 P_1 \cdots P_{t-1})^{-1}.$$

The theorem is proved. ⬜

Similarly, we can prove

THEOREM 9.10

Suppose that scheme $\mathscr{A} = (\mathscr{S}, \mathscr{M}, \mathscr{E}, p_S, p_E)$ *has* $U(t)$-*secrecy* $(1 \leqslant t \leqslant k)$ *and* $p_{S'}$ *is set-uniform. Then its number of encoding rules satisfies*

$$|\mathscr{E}| = \binom{k}{t} \prod_{r=0}^{t-1} P_r^{-1}, \tag{9.18}$$

if and only if the pair

$$(\mathscr{M}, \{\mathscr{M}(e) \mid e \in \mathscr{E}\}) \tag{9.19}$$

is a SPBD t-$(v, b, k; \lambda_1, \lambda_2, \cdots, \lambda_t, 0)$ *with* $\lambda_t = \binom{k}{t}$, *and* p_E, $p_{S'}$ *are uniform distributions. Here*

$$v = |\mathscr{M}|, b = |\mathscr{E}|, k = |\mathscr{S}|, \lambda_r = \binom{k}{t}(P_r P_{r+1}, \cdots P_{t-1})^{-1} \quad (1 \leqslant r \leqslant t-1).$$

Note that if $p_{S'}$ is set-uniform then $p_{S''}$ $(1 \leqslant r \leqslant t-1)$ is also set-uniform.

Suppose that $(\mathscr{M}, \mathscr{F})$ is a SPBD t-$(v, b, k; \lambda_1, \lambda_2, \cdots, \lambda_t, 0)$ and r is a positive integer. Letting each block of $\mathscr{F}$ repeat r times, we form a new k-block family $\mathscr{F}'$. Then $(\mathscr{M}, \mathscr{F}')$ is a new SPBD t-$(v, b', k; \lambda_1', \lambda_2', \cdots, \lambda_t', 0)$ where $b' = rb$, $\lambda_i' = r\lambda_i$ $(1 \leqslant r \leqslant t)$. Obviously, we have

$$\frac{\lambda_{i+1}'}{\lambda_i'} = \frac{\lambda_{i+1}}{\lambda_i} \quad (0 \leqslant i \leqslant t-1),$$

where $\lambda_0 = b$, $\lambda_0' = b'$. Hence a new authentication/secrecy scheme can be constructed by using $(\mathscr{M}, \mathscr{F}')$, which has the same $P_r (0 \leqslant r \leqslant t-1)$ as those of the scheme constructed by using $(\mathscr{M}, \mathscr{F})$. This idea will be used to construct schemes which provide $O(t)$-secrecy or $U(t)$-secrecy, and meet the bound of encoding rules given in Theorem 9.9 or Theorem 9.10.

Recall the SPBD based on rational normal curves (RNC), constructed in Chapter 5. Let $\mathscr{M}$ denote the set of all points in $PG(n, F_q)$ and $\mathscr{F} = \{\mathcal{C}_i \mid 1 \leqslant i \leqslant b\}$ (here b is as given in Theorem 5.1) denote the set of all RNC in $PG(n, F_q)$. Then, the pair $(\mathscr{M}, \mathscr{F})$ is a SPBD with the parameters given in Theorem 5.1. Let $\mathscr{S}$ be the set of all points on the curve $\mathcal{C}$:

$$(1, \alpha, \alpha^2, \cdots, \alpha^n), \alpha \in F_q, \qquad (0, 0, 0, \cdots, 0, 1).$$

Suppose that $\prod_1$ is a $\{t,1\}$-transitive set of permutations on $\mathscr{S}$ which has $k = q + 1$ points, and $\prod_i = \prod_1 (1 \leqslant i \leqslant b)$. Denote the permutations in $\prod_i$ by $\pi_{ij} \{1 \leqslant j \leqslant k!/(k-t)!\}$. For each curve $\mathcal{C}_i$ in $\mathcal{F}$, take one fixed projective transformation T_i which carries $\mathcal{C}$ to $\mathcal{C}_i$, and define encoding rules $e_{ij} \{1 \leqslant j \leqslant k!/(k-t)!\}$ by

$$e_{ij}\big((1,\alpha,\cdots,\alpha^n)\big) = \pi_{ij}\big((1,\alpha,\cdots,\alpha^n)T_i\big),$$
$$e_{ij}\big((0,0,\cdots,0,1)\big) = \pi_{ij}\big((0,0,\cdots,0,1)T_i\big).$$

Thus we obtain an authentication/secrecy scheme which provides $O(t)$-secrecy and has the key minimal property

$$|\mathscr{E}| = \frac{k!}{(k-t)!} \prod_{r=0}^{t-1} P_r^{-1}$$

if p_E and $p_{S'}$ are uniform distributions, by Theorem 9.9.

If we take $\prod$ as a $\{t,1\}$-homogeneous set of permutations on $\mathscr{S}$, instead of a $\{t,1\}$-transitive set of permutations, we obtain a scheme which provides $U(t)$-secrecy with

$$|\mathscr{E}| = \binom{k}{t} \prod_{r=0}^{t-1} P_r^{-1}$$

if p_E and $p_{S'}$ are uniform distributions, by Theorem 9.10.

9.5   Comments

There are various definitions for perfect secrecy. The definitions of $O(t)$-secrecy, $U(t)$-secrecy, and $S(t)$-secrecy adopted in this chapter are followed with Godlewski and Mitchell [13], in which paper they studied the perfect secrecy systems which involve using a key several times ($t \geq 1$), and obtained lower bounds for the number of keys (i.e., encoding rules) in such systems (Theorems 9.1, 9.2) and characters of such systems meeting these lower bounds (Theorems 9.3, 9.5). (The text pp. 187-204 is cited from the text pp. 4-24 of [13] with kind permission of Springer Science and Business Media.) These results for the case when $t = 1$ were first obtained by Shannon [40]. Massey [22], De Soete [9], and Stinson [48] also studied bounds for the number of keys in secrecy systems.

The results of Theorems 9.7 through 9.10 are due to Casse et al.[8]. A construction of perfect authentication/secrecy schemes is also presented in Reference [8]. The presentation of these results in this chapter is a little more explicit than those given in Reference [8]. The construction of perfect authentication/secrecy schemes based on rational normal curves is due to the author.

9.6 Exercises

9.1 Let $\mathscr{S} = \{s_0, s_1, s_2\}$, $\mathscr{E} = \{e_0, e_1, e_2\}$, $\mathscr{M} = \{m_0, m_1, m_2\}$, and suppose that $e_i(s_i) = m_k$ where $k \equiv i + j \ (mod\ 3)$. Suppose also that $p(e_i) = \frac{1}{3}$ for every i. Prove that this scheme provides $U(2)$-secrecy and $O(1)$-secrecy, but not $O(2)$-secrecy.

9.2 For a secrecy scheme, if

$$p(s^r \mid m^r, s_0^{r-1}) = p(s^r \mid s_0^{r-1})$$

for any allowable r-tuple m^r of distinct messages and any r-tuple s^r of distinct source states, where the $(r-1)$-tuple s_0^{r-1} is derived from s^r by deleting its last entry, then the scheme is called secure against order r known-plaintext attack. Prove that if a secrecy scheme provides $O(t)$-secrecy, then it is also secure against order r $(r < t)$ known-plaintext attack.

9.3 Suppose that $\mathscr{A} = (\mathscr{S}, \mathscr{M}, \mathscr{E}, p_S, p_E)$ has $U(t)$-secrecy and $P_i = (k - i)/(v - i)$ for $0 \leq i < t \leq k$, then

$$|\mathscr{E}| \geq \binom{v}{t}.$$

Appendix

A Survey of Constructions for A-Codes

Several constructions for authentication codes, which have not been mentioned in the previous chapters, will be presented in this survey. The notation $k = |\mathscr{S}|$, $v = |\mathscr{M}|$, and $b = |\mathscr{E}|$ are used through the survey.

A.1 Key Grouping Technique

The GMS code (Scheme 3 in Section 3.4) found by Gilbert, MacWilliams, and Sloane [12], was one of the first constructions of authentication codes. For $q = 2$, the GMS code can be represented by encoding Table A.1.

Table A.1

	a	b	c	d	e	f
e_0	0		1			2
e_1	0			1	2	
e_2		0	1		2	
e_3			0	1		2

where 0, 1, and 2 are the three source states, being all points on a given line L, e_0, e_1, e_2, e_3 are the four encoding rules, being other four points not lying on L, and a, b, c, d, e, and f are the messages, being the lines different from L.

In Reference [12] the authors also gave a modified GMS code (Exercise 3.6). For the modified code one chooses, in addition to the line L, also a fixed point P on L. The source states (they are q in number) are then all the points on L except P, the encoding rules are all the points not on L, and the messages are all the lines not passing through P. We obtain a code with q source states, q^2 encoding rules, and q^2 messages. For this code we have $P_0 = P_1 = 1/q$. The modified GMS code for the case $q = 2$ can be illustrated by encoding Table A.2. For this example, we observe that the four encoding rules can be grouped into two groups: $\{e_0, e_3\}$ and $\{e_1, e_2\}$, such that each message contains exactly one encoding rule from each group. This kind of grouping holds in the general case of q. The groups are the lines passing through P except L. For the case $q = 2$, the groups are the lines e and f.

Table A.2

	a	b	c	d	
e_0	0	1			P
e_1	0			1	P
e_2		0	1		P
e_3		0		1	P

THEOREM A.11

In the modified GMS code, if we group the q^2 encoding rules into q groups, such that each group consists of the q rules lying on a line passing through P, then each message contains exactly one encoding rule from each group.

Based on this observation, Smeets, Vanroose, and Wan [46] suggest a construction of codes which provide $U(1)$-secrecy.

In the encoding table of the modified GMS code with $q = 2$, if we interchange 0 and 1 in the rows of the second group of encoding rules $\{e_1, e_2\}$, the encoding Table A.3 of the resulting new code becomes as follows:

Table A.3

	a	b	c	d
e_0	0		1	
e_1	1			0
e_2		1	0	
e_3		0		1

each message contains two source states, one 0 and one 1. Thus, the code provides $U(1)$-secrecy.

For the general case, we label the source states of the modified GMS code with $0, 1, \cdots, q-1$ and the groups of encoding rules also with $0, 1, \cdots, q-1$. We change the encoding table of the modified GMS code according to the following procedure. Replace source state s lying in the row of the encoding rule e and the column of the message m in the encoding table, where e belongs to the group i, by $s-i \pmod{q}$. That is, m will be used to transmit the source state $s - i \pmod{q}$ under the encoding rule e. We can prove:

THEOREM A.12

Each message in the new authentication code contains each of the source states $0, 1, \cdots, q-1$ exactly once, thus the new code provides $U(1)$-secrecy.

The new code still has $P_0 = P_1 = 1/q$.

The above key grouping technique can be also applied to the codes given in Exercise 3.7. Consider all possible lines passing through P, intersecting L only in P, the number of such lines is $Q_n(0,1) - Q_{n-1}(0,1) = q^{n-1}$ {see Lemma 3.6 for the definition of $Q_n(t,r)$}. The encoding rules are grouped such that the encoding rules lying on a common line which intersects L only in P belong to one group. This results in the partitioning of q^n encoding rules into q^{n-1} groups, and each group has q encoding rules. Each message contains $N(n-1,0) - N(n-2,0) = q^{n-1}$ encoding rules {see Lemma 3.5 for the definition of $N(n,t)$}, among them exactly one encoding rule comes from each group. Hence a new authentication code, which provides $U(1)$-secrecy, can be constructed by using the key grouping technique.

A.2   Perpendicular Arrays

If an authentication scheme, with uniform P_E and P_{S^2}, $H(E|S_2, M_2) = 0$ and $P_0 = k/v$, $P_1 = (k-1)/(v-1)$, provides $U(2)$-secrecy, then its number of encoding rules is

$$b = \binom{k}{2} \cdot \frac{v(v-1)}{k(k-1)} = \binom{v}{2}$$

(Theorem 9.8), where k is the number of source states and v is the number of messages.

Stinson [47] describes a construction for authentication schemes with the property that $b = v(v-1)/2$ using perpendicular arrays (Definition 9.4). A perpendicular array $PA(v,k)$ is a $v \cdot (v-1)/2 \times k$ array, A, of the symbols $\{1, \cdots, v\}$, which satisfies the property: for any two columns i and j of A, and for any two distinct symbols x, $y \in \{1, \cdots, v\}$, there is a unique row r such that $\{A(r,i), A(r,j)\} = \{x,y\}$. This is what the condition $H(E|S_2, M_2) = 0$ means.

In the following discussion of this section, we do not assume that P_{S^2} is set-uniform as we did in Theorem 9.8.

THEOREM A.13
If there exists a $PA(v,k)$, where $k > 2$, then there is an authentication code for k source states with v messages and $v(v-1)/2$ encoding rules, which provides $U(2)$-secrecy and has $P_0 = k/v$ when P_E and P_{S_2} are uniform.

PROOF   Regarding the symbols $\{1, \cdots, v\}$ as the set of messages and the columns as the set of source states. For each row $r = (x_1, \cdots, x_k)$ of A, we define an encoding rule e_r such that $e_r(s) = x_s$, $1 \leqslant s \leqslant k$. We check that the code provides $U(2)$-secrecy. For any two messages m, m' and any two source states s and s', there is exactly one encoding rule e such that $\{e(s), e(s')\} = \{m, m'\}$ according to the definition of perpendicular arrays. It implies that

$$p(s, s'|m, m') = p(e) = \frac{2}{v(v-1)} = p(s, s'),$$

by the assumption that P_{S_2} is uniform. Hence, the code provides $U(2)$-secrecy.

Let $n(i, x)$, $1 \leqslant i \leqslant k$, $1 \leqslant x \leqslant v$, denote the number of times the symbol x occurs in the column i. Take any three columns i_1, i_2, and i_3 (since $k > 2$), then we have

$$n(i_1, x) + n(i_2, x) = v - 1,$$
$$n(i_2, x) + n(i_3, x) = v - 1.$$

It follows that $n(i_1, x) = n(i_3, x)$, therefore every symbol occurs exactly $(v - 1)/2$ times in each column. Hence, for any message m, we have

$$|\mathscr{E}(m)| = k \cdot (v - 1)/2$$

and

$$P_0 = \sum_{e \in \mathscr{E}(m)} p(e) = \frac{|\mathscr{E}(m)|}{|\mathscr{E}|} = \frac{k}{v}.$$

☐

The following example illustrates that a scheme constructed by means of Theorem A.13 will not necessarily have $P_1 = (k - 1)/(v - 1)$.

The following is a $PA(5, 3)$:

$$
\begin{array}{ccc}
0 & 1 & 2 \\
1 & 2 & 3 \\
2 & 3 & 4 \\
3 & 4 & 0 \\
4 & 0 & 1 \\
0 & 3 & 1 \\
1 & 4 & 2 \\
2 & 0 & 3 \\
3 & 1 & 4 \\
4 & 2 & 0 \\
\end{array}
$$

There are ten encoding rules e_i $(1 \leqslant i \leqslant 10)$ in the scheme constructed by the perpendicular array. Suppose the source probability distribution is (p_1, p_2, p_3), where $p_1 > p_2 > p_3$. If the opponent observes the message 0, then we have the conditional probability distribution on the encoding rules:

$$p(e|M = 0) = \frac{p(e, M = 0)}{p(M = 0)} = \frac{p(e)p(f_e(0))}{\sum\limits_{e' \in \mathscr{E}(0)} p(e')p(f_{e'}(0))}$$

$$= \frac{p(f_e(0))}{2 \sum\limits_{i=1}^{3} p_i} = \frac{p(f_e(0))}{2}.$$

If the opponent substitutes message 0 with message 1, it will be accepted as authentic with the probability

$$\frac{p(f_{e_1}(0))}{2} + \frac{p(f_{e_5}(0))}{2} + \frac{p(f_{e_6}(0))}{2} = \frac{p_1}{2} + \frac{p_2}{2} + \frac{p_1}{2} = \frac{1}{2} + \frac{p_1 - p_3}{2}.$$

In fact, the opponent's optimal substitution strategy is to replace any message m by the message $(m+1)$ $(mod\ 5)$. This yields

$$P_1 = \frac{1}{2} + \frac{p_1 - p_3}{2} \neq \frac{1}{2}.$$

A special type of perpendicular array will allow us to attain $P_1 = (k - 1)/(v - 1)$. A $PA(v, k)$ is said to be cyclic {and is denoted as $CPA(v, k)$} if any cycle permutation of the columns of A yields an array which can be obtained from A by means of a suitable permutation of the rows of A. That is, if $(x_1, \cdots, x_k)$ is a row of A, then $(x_2, \cdots, x_k, x_1)$ is also a row of A.

The following is a $CPA(5, 5)$:

$$
\begin{array}{ccccc}
0 & 1 & 2 & 3 & 4 \\
1 & 2 & 3 & 4 & 0 \\
2 & 3 & 4 & 0 & 1 \\
3 & 4 & 0 & 1 & 2 \\
4 & 0 & 1 & 2 & 3 \\
1 & 3 & 0 & 2 & 4 \\
3 & 0 & 2 & 4 & 1 \\
0 & 2 & 4 & 1 & 3 \\
2 & 4 & 1 & 3 & 0 \\
4 & 1 & 3 & 0 & 2 \\
\end{array}
$$

THEOREM A.14

If there exists a $CPA(v, k)$, then there is an authentication/secrecy code with k source states and v messages which provides $U(2)$-secrecy and has $P_0 = k/v$, $P_1 = (k - 1)/(v - 1)$ when P_E and P_{S_2} are uniform (i.e., it is 2-fold perfect with combinatorial bounds).

PROOF Let A be a $CPA(v, k)$. Construct a code as in Theorem A.13. We need only to prove that $P_1 = (k - 1)/(v - 1)$. Let m and m' be any two distinct messages. There are $k \cdot (k - 1)/2$ rows r of A for which m, m' occur in row r. For each source state j, there are exactly $(k - 1)/2$ encoding rules e_r where m, m' occur in row r and $e_r(j) = m'$, since A is cyclic. Hence,

$$P(m|m') = \sum_{e \in \mathscr{E}(m,m')} p(e|M = m') = \sum_{e \in \mathscr{E}(m,m')} \frac{p(e, M = m')}{p(M = m')}$$

$$= \frac{\sum\limits_{e \in \mathscr{E}(m,m')} p(e)p(f_e(m'))}{\sum\limits_{e \in \mathscr{E}(m')} p(e)p(f_e(m'))} = \frac{(k-1)/2 \cdot \sum\limits_{j=1}^{k} p(S=j)}{(v-1)/2 \cdot \sum\limits_{j=1}^{k} p(S=j)} = \frac{k-1}{v-1},$$

where $P(m|m')$ is the probability that the message m is accepted as authentic given that m' has been observed. Thus, the desired result is followed.   ◻

It is not difficult to see that the existence of a $CPA(v,k)$ requires that $(v-1)/2$ and $(k-1)/2$ be integers, and that $k|v(v-1)/2$, i.e., that v and k be odd, and that $2k|v(v-1)$. For $k = 3$ and 5, these conditions are necessary and sufficient for existence, with one exception:

THEOREM A.15

1. A $CPA(v,3)$ exists if and only if $v \equiv 1$ or 3 modulo 6 [21].

2. A $CPA(v,5)$ exists if and only if $v \equiv 1$ or 5 modulo 10, $v \neq 15$ [20].

For $k > 5$, only sporadic results are known. One class of CPA is given by

PROPOSITION A.2

Suppose that k is odd and $v \equiv 1 \pmod{2k}$ is a prime power, then there exists a $CPA(v,k)$.

PROOF   Let w be a primitive element in the finite field $\mathbb{F}_v$, and let $\alpha = w^{(v-1)/k}$. For each $i = 1, \cdots, (v-1)/2k$, for each $j = 0, \cdots, k-1$, and, for each $\beta \in \mathbb{F}_v$, define a row

$$\beta + w^i\alpha^j, \ \beta + w^i\alpha^{1+j}, \ \beta + w^i\alpha^{2+j}, \ \cdots, \ \beta + w^i\alpha^{k-1+j}.$$

This defines $v \cdot (v-1)/2$ rows, which is the right number at least. In the set of numbers

$$\{i + j \cdot (v-1)/k\}, \quad i = 1, \cdots, (v-1)/2k, \ j = 0 \cdots, k-1,$$

there are no two numbers with difference $(v-1)/2$. So in the set of elements

$$I = \{w^i\alpha^j\}, \quad i = 1, \cdots, (v-1)/2k, \ j = 0, \cdots, k-1,$$

there are no two elements γ, γ' such that $\gamma = -\gamma'$. The number of elements in the set I is $(v-1)/2$. For any two distinct elements a and b in $\mathbb{F}_v$, and for any s_1 and s_2 ($0 \leqslant s_1 < s_2 \leqslant k-1$), one and only one element from $(a-b)/(\alpha^{s_1} - \alpha^{s_2})$ and $(b-a)/(\alpha^{s_1} - \alpha^{s_2})$ belongs to I, say $(a-b)/(\alpha^{s_1} - \alpha^{s_2}) = w^i\alpha^j \in I$, then

$$a - b = w^i \alpha^{s_1+j} - w^i \alpha^{s_2+j}.$$

Put

$$\beta = a - w^i \alpha^{s_1+j} = b - w^i \alpha^{s_2+j}.$$

This proves the desired result. $\quad\square$

Hence, we can pick a random encoding rule e by generating a random 3-tuple (i, j, β) of the form described above. A source state s $(0 \leqslant s \leqslant k - 1)$ would then be encoded as

$$e(s) = \beta + w^i \alpha^{s+j}.$$

Also, given a message m, we can solve the source state s by means of the equation

$$\alpha^{s+j} = (m - \beta)w^{-i}.$$

This requires the calculation of a logarithm in the finite field $\mathbb{F}_v$.

The text pp. 217-221 of this section is cited from the text pp. 124-126 of [47] with kind permission of Springer Science and Business Media.

A.3   Generalized Quadrangles

A generalized quadrangle (GQ) is an incidence structure $Q = (P, L, I)$ in which P and L are disjoint (nonempty) sets of objects called points and lines, respectively, and for which I is a symmetric point-line incidence relation satisfying the following axioms:

1. Each point is incident with $1 + t$ lines $(t \geqslant 1)$ and two distinct points are incident with at most one line.

2. Each line is incident with $1 + s$ points $(s \geqslant 1)$ and two distinct lines are incident with at most one point.

3. If x is a point and l a line not incident with x, then there exists a unique line l' and a unique point p' such that $x I l' I p' I l$.

The integers s and t are the parameters of the GQ and Q is said to have order (s, t). There is a point-line duality for GQ {of order (s, t)} for which in any definition or theorem the words "point" and "line" are interchanged and the parameters s and t are interchanged.

Example A.1

The simplest example of generalized quadrangles is the quadrangle, i.e., the incidence structure consisting of four points p_1, p_2, p_3, and p_4, no three of which are collinear, and four lines p_1p_2, p_2p_3, p_3p_4, and p_4p_1, where by $p_i p_j$ we mean the line incident with both p_i and p_j. More precisely, let $P = \{p_1, p_2, p_3, p_4\}$, $L = \{l_1, l_2, l_3, l_4\}$, and $I = \{(p_i, l_i), (p_{i+1}, l_i) \mid i = 1, 2, 3, 4 \text{ and } p_5 = p_1\}$, then (P, L, I) is called a quadrangle and is clearly a generalized quadrangle. ☐

LEMMA A.15

Let v and b denote the number of points and the number of lines in a GQ, respectively. In any generalized quadrangle with parameters s and t, $v = (s+1)(st+1)$ and $b = (t+1)(st+1)$.

PROOF   Fix a line l. There are $v - (s+1)$ points not on l. By the axiom 3 each point not on l lies on a unique line meeting l. Hence, an alternate way to count all the points not on l is to count all the points on the lines meeting l but not lying on l. There are $s + 1$ points on l, through each of these $s + 1$ points there pass t lines different from l, and on each of these $(s + 1)t$ lines there are s points not lying on l. Therefore, altogether there are $(s + 1)st$ points not on l. Hence, $v - (s+1) = (s+1)ts$, or $v = (s+1)(ts+1)$. Dually $b = (t+1)(st+1)$. ☐

Let $p, q \in P$. We write $p \sim q$ and say that p and q are collinear, provided that there is some line l for which $pIlIq$. And $p \nsim q$ means that p and q are not collinear. We agree that $p \sim p$ for every $p \in P$.

For any point p of a generalized quadrangle $Q = (P, L, I)$, define

$$p^{\perp} = \{x \in P \mid x \sim p\}.$$

For $p, q \in P$ and $p \neq q$, the trace of p and q is defined by

$$\text{tr}(p, q) = \{p, q\}^{\perp} = \{x \in P \mid x \sim p \text{ and } x \sim q\}.$$

LEMMA A.16

In a generalized quadrangle $Q = (P, L, I)$ with parameters s and t, we have

$$|p^{\perp}| = st + s + 1 \qquad \text{for any } p \in P,$$

$$|\text{tr}(p, q)| = \begin{cases} s + 1 & \text{for any } p, q \in P \text{ such that } p \sim q \text{ and } p \neq q, \\ t + 1 & \text{for any } p, q \in P \text{ such that } p \nsim q. \end{cases}$$

PROOF   Let $p \in P$. The number of lines passing through p is equal to $t + 1$, and on each of these lines there are s points different from p. Clearly, this gives us precisely $(t + 1)s$ points which are incident with p and different from p. With the point p added, we obtain $|p^{\perp}| = st + s + 1$.

Let $p, q \in P$. If $p \sim q$ and $p \neq q$, then $\text{tr}(p, q)$ is the set of points on the line pq, and therefore, $|\text{tr}(p, q)| = s + 1$. Now suppose that $p \nsim q$. There are

$t + 1$ lines passing through p. By the axiom 3 there is one and only one point on each of these lines, which is collinear with q. Clearly, this gives precisely $t + 1$ points which are collinear with both p and q, i.e., $|\text{tr}(p, q)| = t + 1$. □

We construct an authentication code from a generalized quadrangle with parameters s and t as follows: take a fixed point p. Let the source states be the $t + 1$ lines which pass through p, the encoding rules be the points not collinear with p, and the messages be the points of $p^{\perp} \setminus \{p\}$. Given a source state l that is a line passing through p and an encoding rule $e \notin p^{\perp}$, by the axiom 3 there is a unique point m lying on l and being collinear with e. Then we say that the source state l is encoded by e into the message m.

THEOREM A.16
If there exists a GQ with parameters (s, t), then there is a Cartesian authentication code with $|\mathscr{S}| = t + 1$, $|\mathscr{M}| = (t + 1)s$, $|\mathscr{E}| = ts^2$, and $P_0 = P_1 = 1/s$.

PROOF Consider the authentication code defined above. It is easy to see that $|\mathscr{S}| = t + 1$, $|\mathscr{M}| = (t + 1)s$ and

$$|\mathscr{E}| = (s + 1)(st + 1) - (t + 1)s - 1 = s^2 t.$$

This is a Cartesian code, since each message m is always used to transmit a unique source state, which is the line passing through m and p.

Assume that the encoding rules are equally probable, i.e., each one has a probability of occurrence $1/s^2 t$. For an arbitrary message m, suppose that m lies on the line l passing through p. Then there are t lines passing through m and distinct from l and on each of these t lines there are s points distinct from m. These st points are all the encoding rules which will encode l into m. Therefore, $|\mathscr{E}(m)| = st$ and $P_0 = st/s^2 t = 1/s$.

For any two distinct messages m and m', if $m \sim m'$ then the line passing through m and m' is a source state s, hence $(m, m') \subset \mathscr{M}(s)$ and $\mathscr{E}(m, m') = \emptyset$. Now we assume that $m \not\sim m'$, then we have

$$\mathscr{E}(m, m') = \{e \in \mathscr{E} \,|\, e \sim m, \, e \sim m', \, e \neq p\}$$
$$= \text{tr}(m, m') \setminus \{p\}.$$

Therefore, $|\mathscr{E}(m, m')| = t$ by Lemma A.16. It implies that

$$P_1 = \frac{|\mathscr{E}(m, m')|}{|\mathscr{E}(m)|} = \frac{t}{st} = \frac{1}{s}.$$

□

Let l_i $(1 \leqslant i \leqslant t + 1)$ be the $t + 1$ lines incident with p. For an arbitrary encoding rule e defined in the preceding theorem, suppose that $e(l_i) = m_i$ which is a point incident with l_i. We define $t + 1$ new encoding rules by

$$e^{(j)}(l_i) = m_{i+j \pmod{t+1}}, \ 1 \le j \le t+1.$$

Thus we obtain a new code with

$$|\mathscr{S}| = t+1, \ |\mathscr{M}| = (t+1)s, \ |\mathscr{E}| = ts^2(t+1).$$

It is easy to check that $P_0 = P_1 = 1/s$ for the new code. Besides, the new code provides $U(1)$-secrecy. By observing we find that $\#\{e \in \mathscr{E} \mid e(l) = m\} = st$ for arbitrary source state l and arbitrary message m; hence

$$p(l|m) = \frac{p(l,m)}{p(m)} = \frac{p(l) \cdot \sum\limits_{e:\,e(l)=m} p(e)}{\sum\limits_{e \in \mathscr{E}} p(e)p(f_e(m))}$$

$$= \frac{p(l) \cdot \#\{e \in \mathscr{E} \mid e(l) = m\}}{\sum\limits_{i=1}^{t+1} p(l_i) \cdot \#\{e \in \mathscr{E} \mid e(l_i) = m\}}$$

$$= p(l).$$

A spread of a generalized quadrangle Q is a set R of lines of Q such that each point of Q is incident with a unique line of R. Hence, there are $st+1$ lines in R.

Consider a generalized quadrangle Q of order (s,t) which contains a spread $R = \{l_1, l_2, \cdots, l_{st+1}\}$. Define the source states as the lines of R ($k = st + 1$) and the messages as the points of Q $\{v = (s+1)(st+1)\}$. Denote the points incident with l_i as $x_{i,1}, x_{i,2}, \cdots, x_{i,s+1}$ ($1 \le i \le st+1$). We associate with each point x_{ij} an encoding rule

$$e_{i,j}(l_n) = x_{i+n,h}, \quad 1 \le n \le st+1$$

with $x_{i+n,h}$ the unique point on the line l_{i+n} which is collinear with $x_{i,j}$ (where $i+n$ is taken modulo $st+1$). In this way we obtain $|\mathscr{E}| = (1+s)(st+1)$.

THEOREM A.17
If there exists a GQ with parameters (s,t) containing a spread R, then there is an authentication code with $P_0 = k/v$ which provides $U(1)$-secrecy.

PROOF Consider the code defined above. Assume that each encoding rule is used with a probability $1/(s+1)(st+1)$. For arbitrary message $m = x_{i_0,j_0}$, we have $|\mathscr{E}(m)| = st + 1$ {since there are st points which are collinear with x_{i_0,j_0}, not lying on the line of the spread incident with x_{i_0,j_0}, and x_{i_0,j_0} itself also determines an encoding rule of $\mathscr{E}(x_{i_0,j_0})$}. Hence,

$$P_0 = \frac{st+1}{(s+1)(st+1)} = \frac{1}{s+1} = \frac{k}{v}.$$

If $e_{i,j}(l_n) = x_{i_0,j_0}$, then $i + n \equiv i_0 \pmod{st + 1}$ and when n runs through $1, 2, \cdots, st + 1$, so does i. This implies that $\#\{e \in \mathscr{E} \mid e(l) = m\} = 1$ for arbitrary l and m, hence,

$$
\begin{aligned}
p(l|m) &= \frac{p(l, m)}{p(m)} \\
&= \frac{p(l) \cdot \sum\limits_{e : e(l) = m} p(e)}{\sum\limits_{e} p(e)p(f_e(m))} \\
&= \frac{p(l)}{\sum\limits_{i=1}^{st+1} p(l_i)} = p(l)
\end{aligned}
$$

The theorem is proved. ☐

For the known GQ we refer to Reference [27] or [52]. Here we introduce an example of the known GQ.

Consider the projective space $PG(3, \mathbb{F}_q)$ which arises from the vector space $\mathbb{F}_q^{(4)}$ by regarding the 1-dimensional vector subspaces as the points and the $(r+1)$-dimensional vector subspaces as the projective r-dimensional subspaces $(0 \leqslant r \leqslant 3)$. Define a 4×4 nonsingular alternate matrix:

$$
K = \begin{pmatrix} 0 & I^{(2)} \\ -I^{(2)} & 0 \end{pmatrix}.
$$

Let P be an r-dimensional subspace of $PG(3, \mathbb{F}_q)$ and the $(r + 1)$-dimensional subspace of $\mathbb{F}_q^{(4)}$ corresponding to P will also be denoted by P. The subspace P is called totally isotropic with respect to K if $PKP^{\perp} = 0$. We know from Theorem 8.3 that the maximal totally isotropic subspaces of $PG(3, \mathbb{F}_q)$ are of dimension 1. All points of $PG(3, \mathbb{F}_q)$ are totally isotropic of 0-dimension. Define

$$
P^{\perp} = \{p \in PG(3, \mathbb{F}_q) \mid pKP^t = 0\}.
$$

THEOREM A.18

The set of points of $PG(3, \mathbb{F}_q)$ together with the set of totally isotropic lines with respect to the nonsingular alternate matrix K forms a generalized quadrangle.

PROOF   It is necessary to show that three axioms of GQ are satisfied.

1.   We will show that each point is incident with the same number of totally isotropic lines. Without loss of generality we consider only the case

$p = (1, 0, 0, 0)$. Suppose that $p' = (0, x_1, x_2, x_3)$ and

$$\binom{p}{p'} K \binom{p}{p'}^t = 0.$$

It implies that $x_2 = 0$ and $p' = (0, 1, 0, x)$ $(x \in \mathbb{F}_q)$ or $p' = (0, 0, 0, 1)$. Hence each point is incident with $q + 1$ totally isotropic lines. It is trivial that two distinct points are incident with at most one totally isotropic line.

2. It is trivial that each totally isotropic lines is incident with $q + 1$ points and two distinct totally isotropic lines are incident with at most one point.

3. Given a totally isotropic line P and a point $p \notin P$, then $P \nsubseteq p^{\perp}$; in fact if $P \subset p^{\perp}$, then $P = P^{\perp} \supset (p^{\perp})^{\perp} = p$, a contradiction. Hence, $p' = P \cap p^{\perp}$ is a point on P, which is the unique point on P such that the line incident with both p' and p is totally isotropic.   ⬚

The text pp. 223-225 of this section is cited from the text pp. 66-69 of [9] with kind permission of Springer Science and Business Media.

A.4   Resolvable Block Design and A^2-Codes

DEFINITION A.7   *A block design (V, B) is called α-resolvable if the block set B can be partitioned into classes $C_1, C_2, \cdots, C_k$ with the property that in each class, every point of V occurs in exactly α blocks.*

Obana and Kurosawa [26] and Wang, Safavi-Naini, and Pei [55] studied the application of α-resolvable block designs in construction of perfect Cartesian A^2-codes. The following result was proved in Reference [55], but it will be explained with some modification by using the results of Chapter 4.

Assume that $\mathscr{A} = (\mathscr{S}, \mathscr{M}, \mathscr{E}_R, \mathscr{E}_T, p_S, p_{E_R}, p_{E_T})$ is a t-fold perfect Cartesian A^2-code of Type I, i.e., for any t-subset $s_{i_1}, s_{i_2}, \cdots, s_{i_t}$ of $\mathscr{S}$ and any $m^t \in \mathscr{M}_{i_1, i_2, \cdots, i_t}$, we have $|\mathscr{E}_T(m^t)| = 1$ as we assumed in Corollary 4.7. Consider the pair

$$(V, B) = (\mathscr{E}_R, \{\mathscr{E}_R(m) \mid m \in \mathscr{M}\});$$

put

$$C_i = \{\mathscr{E}_R(m) \mid m \in \mathscr{M}(s_i)\}, \quad 1 \leqslant i \leqslant k.$$

Each class C_i contains $l = v/k$ blocks (Corollary 4.6), and the block set B is partitioned into the classes $C_1, C_2, \cdots, C_k$. For any decoding rule $f \in V$ and any source state s, let $\mathscr{M}(f, s) = \{m \in \mathscr{M} \mid f(m) = s\}$ as defined in Chapter

2. We know that $|\mathcal{M}(f,s)| = c$ is a constant (Corollary 4.5). This means that the pair (V, B) is c-resolvable.

THEOREM A.19

The pair $(V, B) = (\mathcal{E}_R, \{\mathcal{E}_R[m] \mid m \in \mathcal{M}\})$ is a c-resolvable block design with the following properties:

(1) Any collection of r $(1 \leqslant r \leqslant t)$ blocks from r different classes intersect in $\mu_r (> 0)$ points {where $\mu_r = (\frac{l}{t})^{t-r} \frac{(l-1)}{(c-1)}$}

(2) For any t blocks $B_{j_1}, B_{j_2}, \cdots, B_{j_t}$ from different classes $C_{j_1}, C_{j_2}, \cdots, C_{j_t}$ and for any $u\, (\neq j_1, j_2, \cdots, j_t)$, there exists a unique block $B_u \in C_u$ such that

$$B_{j_1} \cap B_{j_2} \cap \cdots \cap B_{j_t} = B_{j_1} \cap B_{j_2} \cap \cdots \cap B_{j_t} \cap B_{j_u}.$$

Furthermore, for any $B \in C_u \setminus \{B_u\}$, $|B_{j_1} \cap B_{j_2} \cap \cdots \cap B_{j_t} \cap B| = 1$.

PROOF We have showed above that (V, B) is c-resolvable. Let $m^r = (m_1, m_2, \cdots, m_r) \in \overline{\mathcal{M}_R^r}$, then

$$\mathcal{E}_R(m^r) = \mathcal{E}_R(m_1) \cap \cdots \cap \mathcal{E}_R(m_r).$$

The property (1) is deduced from (iii), (v), and (vi) of Corollary 4.7.

Consider the proof of the property (2). Let $B_{j_r} = \mathcal{E}_R(m_{j_r})$ $(1 \leqslant r \leqslant t)$, then $B_{j_1} \cap B_{j_2} \cdots \cap B_{j_t} \neq \emptyset$ by (1). Suppose $f \in B_{j_1} \cap B_{j_2} \cap \cdots \cap B_{j_t}$. There exists a unique encoding rule e such that $(m_{j_1}, m_{j_2}, \cdots, m_{j_t}) \subset \mathcal{M}(e) \subset \mathcal{M}(f)$ {Corollary 4.7, (iii)}. For any $u\, (\neq j_1, \cdots, j_t)$, let $e(s_u) = m_u$, then $f \in \mathcal{E}_R(m_u)$, therefore $f \in B_{j_1} \cap \cdots \cap B_{j_t} \cap B_u$ where $B_u = \mathcal{E}_R(m_u)$. This implies

$$B_{j_1} \cap \cdots \cap B_{j_t} = B_{j_1} \cap \cdots \cap B_{j_t} \cap B_u.$$

Let $m \in \mathcal{M}(s_u)$, $m \neq m_u$, then there exists a unique decoding rule f such that $(m_{j_1}, \cdots, m_{j_t}, m) \subset \mathcal{M}(f)$ {Corollary 4.7 (viii)}. It follows that $|B_{j_1} \cap \cdots \cap B_{j_t} \cap B| = 1$ where $B = \mathcal{E}_R(m) \in C_u \setminus \{B_u\}$. □

Now we assume that the pair (V, B) is an arbitrary α-resolvable block design with the properties (1) and (2) of Theorem A.19 (the expression for μ_r is not needed), the block set B is partitioned into classes $C_1, C_2, \cdots, C_k$, and all classes have the same number of blocks, i.e., $|C_i| = l$ for $1 \leqslant i \leqslant k$. We regard each block B_i in B as a message and it is still denoted by B_i. Let U be the set of $v = kl$ messages. Fix t classes, say $C_1, C_2, \cdots, C_t$, for any t blocks $B_1 \in C_1, B_2 \in C_2, \cdots, B_t \in C_t$ and any u $(t < u \leqslant k)$. There exists a unique block $B_u \in C_u$ such that $B_1 \cap \cdots \cap B_t = B_1 \cap \cdots \cap B_t \cap B_u$ by the property (2). In this case there exists only one group $B_{t+1} \in C_{t+1}, \cdots, B_k \in C_k$ such that

$$B_1 \cap \cdots \cap B_t = B_1 \cap \cdots \cap B_t \cap B_{t+1} \cap \cdots \cap B_k.$$

Now we define a block (k-subset) of U,

$$(B_1, B_2, \cdots, B_k). \tag{A.20}$$

Thus, we get l^t blocks which form a block set E.

THEOREM A.20
Suppose that there exists an α-resolvable design (V, B) where B is partitioned into classes $C_1, C_2, \cdots, C_k$ with the properties (1) and (2) of Theorem A.19 and all classes have the same number of blocks, i.e., $|C_i| = l$ for $1 \leqslant i \leqslant k$. Then, there exists a t-fold perfect Cartesian A^2-code, which has the following parameters:
(i) *The number of source states is k;*
(ii) *The number of messages is $|B| = lk$;*
(iii) *$|\mathscr{E}_R| = |V| = lk$, $|\mathscr{E}_T| = l^t$;*
(iv) *$P_{O_r} = \alpha/l$, $P_{R_i} = 1/\alpha$, $P_T = (\alpha - 1)/(l - 1)$, $0 \leqslant i \leqslant t - 1$.*

PROOF   Let the sets U and E be defined as above. We regard the set U as the set of messages, and regard each class C_i as a source state s_i, thus, the set of source states is $\mathscr{S} = \{s_1, s_2, \cdots, s_k\}$. Hence (i) and (ii) hold. For each point $f \in V$ define a decoding rule which is also denoted by f, such that

$$f(B_i) = \begin{cases} s_i & \text{if } f \in B_i \in C_i \\ \text{reject} & \text{if } f \notin B_i. \end{cases}$$

The definition of decoding rules shows that $|\mathscr{M}(s)| = l$ and $|\mathscr{M}(f,s)| = \alpha$ for arbitrary decoding rule f and arbitrary source states s.

For each point in Equation (A.20) of E define an encoding rule e such that

$$e(s_i) = B_i, \quad 1 \leqslant i \leqslant k.$$

Thus, the set V is the set of decoding rules and the set E is the set of encoding rules, and $|\mathscr{E}_R| = |V| = lk$ and $|\mathscr{E}_T| = |E| = l^t$. Hence, (iii) is proved. This A^2-code is Cartesian of Type I since the assumption of Corollary 4.7 is satisfied.

Given an encoding rule e and a decoding rule f if and only if

$$f \in B_1 \cap B_2 \cap \cdots \cap B_k$$

where $B_i = e(s_i)$ for $1 \leqslant i \leqslant k$, e is valid under f.

Assume that $\mathscr{E}_R$, $\mathscr{E}_T$, and $\mathscr{E}_R(e)$ for any e {therefore, $\mathscr{E}_T(f)$ for any f} have uniform probability distribution. For any r-subset $(1 \leqslant r \leqslant t)$ $m^r = \{B_{i_1}, B_{i_2}, \cdots, B_{i_r}\}$, if $B_{i_j} \in C_{i_j}$ for $1 \leqslant j \leqslant r$, then

$$|\mathscr{E}_R(m^r)| = \#\{f \mid f \in B_{i_1} \cap B_{i_2} \cap \cdots \cap B_{i_r}\} = \mu_r$$

by the property (1), otherwise $|\mathscr{E}(m^r)| = 0$. Hence,

$$P_{O_0} = \frac{\mu_1}{|\mathscr{E}_R|}, \quad P_{O_r} = \frac{\mu_{r+1}}{\mu_r} \quad 1 \leqslant r \leqslant t - 1 \quad (A.21)$$

according to Corollary 4.1.

Given any r-subset ($0 \leqslant r \leqslant t$) of message $m^r = \{B_{i_1}, B_{i_2}, \cdots, B_{i_r}\}$ such that $B_{i_j} \in C_{i_j}$ for $1 \leqslant j \leqslant r$ and $f \in B_{i_1} \cap B_{i_2} \cap \cdots \cap B_{i_r}$. Since f occurs in exactly α blocks in each class $C_{i_{r+1}}, \cdots, C_{i_t}$, there are α^{t-r} choices of $\{B_{i_{r+1}}, \cdots, B_{i_t}\}$ with $B_{i_j} \in C_{i_j}$ for $r + 1 \leqslant j \leqslant t$ such that

$$f \in B_{i_1} \cap \cdots \cap B_{i_r} \cap B_{i_{r+1}} \cap \cdots \cap B_{i_t}.$$

Hence $|\mathscr{E}_T(f, m^r)| = \alpha^{t-r}$ according to the property (2) and the definition of encoding rules, it implies that

$$P_{R_r} = \frac{|\mathscr{E}_T(f, m^{r+1})|}{|\mathscr{E}_T(f, m^r)|} = \frac{1}{\alpha}, \quad 0 \leqslant r \leqslant t - 1 \quad (A.22)$$

by Corollary 4.2.

For a given encoding rule e and a message $m' \in \mathscr{M}'(e)$, we may assume that $m' \in C_u$ with $t < u \leqslant k$. Hence $|\mathscr{E}_R(e, m')| = 1$ by the property (2) and

$$P_T = \frac{|\mathscr{E}_R(e, m')|}{|\mathscr{E}_R(e)|} = \frac{1}{\mu_t} \quad (A.23)$$

according to Corollary 4.3.

Using the proof of the items (v) and (vi) of Corollary 4.7 we obtain

$$P_{O_r} = \alpha/l, \quad 0 \leqslant r \leqslant t - 1,$$

and

$$P_T = (\alpha - 1)/(l - 1),$$

thus, (4) holds. Furthermore, by Equations (A.21) and (A.23) we obtain

$$\mu_r = P_{O_{r+1}} \cdots P_{O_t} P_T^{-1} = \left(\frac{\alpha}{l}\right)^{t-r} \cdot \frac{l-1}{\alpha-1}, \quad 1 \leqslant r \leqslant t.$$

Finally, it follows that

$$|\mathscr{E}_R| = |V| = (P_{O_0} P_{O_1} \cdots P_{O_{t-1}} P_T)^{-1},$$
$$|\mathscr{E}_T| = l^t = (P_{O_0} \cdots P_{O_{t-1}} P_{R_0} \cdots P_{R_{t-1}})^{-1},$$

the code constructed is a t-fold perfect A^2-code. $\qquad\qquad\square$

A.5   Regular Bipartite Graphs

Suppose that the scheme $\mathscr{A} = (\mathscr{S}, \mathscr{M}, \mathscr{E}, p_S, p_E)$ with $P_0 = k/v$ is $U(1)$-secrecy and p_S is uniform, then

$$|\mathscr{E}| \geqslant \binom{k}{1} \cdot \frac{v}{k} = v$$

(Theorem 9.8). The equality holds if and only if the pair

$$(\mathscr{M}, \{\mathscr{M}(e) \mid e \in \mathscr{E}\})$$

is a 1-(v, k, k) design and p_E is uniform (Theorem 9.10). A construction of this kind of code by using k-regular bipartite graphs was suggested by Feng and Kwak [10].

Construct a graph G taking the set of vertices as

$$V(G) = \mathscr{E} \cup \mathscr{M},$$

and the set of edges as

$$E(G) = \{\{e, m\} \mid m \in \mathscr{M}, e \in \mathscr{E}(m)\}.$$

Since $|\mathscr{S}| = k$, $|\mathscr{E}(m)| = k$ for any $m \in \mathscr{M}$, it is clear that G is a k-regular bipartite graph with bipartition $\mathscr{E}$ and $\mathscr{M}$.

Conversely, suppose that G is a k-regular bipartite graph with bipartition (X, Y). The graph G is k-edge-colorable [19]. Let G be edge-colored by k different colors, say $s_1, s_2, \cdots, s_k$, thus the k-edges coming from an arbitrary vertex are colored by k different colors, respectively. Set $\mathscr{S} = \{s_1, s_2, \cdots, s_k\}$, $\mathscr{E} = X$, and $\mathscr{M} = Y$. Define an encoding rule e by $e(s) = m$ for any $s \in \mathscr{S}$ and $e \in \mathscr{E}$ where $\{e, m\}$ is an edge of G colored by s. Then $|\mathscr{E}(m)| = k$ for any $m \in \mathscr{M}$, the code constructed is desired.

Example A.2
Let A be an alphabet of q symbols $\{0, 1, \cdots, q - 1\}$ and let X be the set of all n-tuples of elements of A. Construct a bipartite graph having bipartition (X, X) where $\{x, y\}$ $(x, y \in X)$ is an edge if and only if x and y differ in just 1 coordinate. The graph is $n(q - 1)$-regular bipartite with $2q^n$ vertices. An authentication code with $|\mathscr{S}| = n(q - 1)$, $|\mathscr{M}| = |\mathscr{E}| = q^n$, $P_0 = n(q - 1)/q^n$, and $U(1)$-secrecy can be constructed using the graph. ⬜

Example A.3
Let G be a group of order n and let S be a k-subset of G. Set $\mathscr{S} = S$, $\mathscr{E} = \mathscr{M} = G$. Define an encoding rule e by $e(s) = es$ for any $s \in \mathscr{S}$ and $e \in \mathscr{E}$. For any message m and any source state s, there is an encoding

rule $e = s^{-1}m$ such that $e(s) = m$, hence $|\mathscr{E}(m)| = n$ for any message m. An authentication code with $|\mathscr{S}| = k$, $|\mathscr{E}| = |\mathscr{M}| = n$, $P_0 = k/n$ and $U(1)$-secrecy can be constructed by the group. $\square$

References

[1] Beth, T., Jungnichel, D., and Lenz, H. *Design Theory, Volume 1 and 2 (Second Edition)*, Cambridge University Press, Cambridge, 1999.

[2] Bonisoli, A. and Quattrocchi, P. Existence and extension of sharply k-transitive permutation sets: a survey and some new results, *Ars Combinatoria*, 24A, 163, 1987.

[3] Bose, R.C. On the construction of balanced incomplete block designs, *Annals of Eugenics*, 9, 353, 1939.

[4] Bose, B.C. and Shrikhande, S.S. On the construction of sets of mutually orthogonal Latin squares and the falsity of a conjecture of Euler, *Transactions of the American Mathematical Society*, 95, 191, 1960.

[5] Bose, R.C., Shrikhande, S.S., and Parker, E.T. Further result on the construction of mutually orthogonal Latin squares and the falsity of Euler's conjecture, *Canadian Journal of Mathematics*, 12, 189, 1960.

[6] Brickell, E.F. A new result in message authentication, *Congressus Numerantium*, 43, 141, 1984.

[7] Bush, K.A. Orthogonal arrays of index unity, *Annals of Mathematical Statistics*, 23, 426, 1952.

[8] Casse, L.R.A., Martin, K.M., and Wild, P.R. Bounds and characterization of authentication/secrecy schemes, *Design, Codes and Cryptograph*, 13, 107, 1998.

[9] De Soete, M. Some constructions for authentication-secrecy codes, *Advances in Cryptology–Eurocrypt'88, Lecture Notes in Computer Science 330*, Springer-Verlag, Berlin, 57, 1988.

[10] Feng, R. and Kwak, J.H. Minimal authentication codes with perfect secrecy, *Research Report 34*, Institute of Mathematics and School of Mathematical Sciences, Peking University, 2000.

[11] Feng, R. and Wan, Z. A construction of Cartesian authentication codes from geometry of classical groups, *Journal of Combinatorics, Information and System Sciences*, 20, 197, 1995.

[12] Gilbert, E.N., Macwilliams, F.J., and Sloane, N.J.A. Codes which detect deception, *Bell System Technical Journal*, 53, 405, 1974.

[13] Godlewski, P. and Mitchell, C. Key-minimal cryptosystems for unconditional secrecy, *Journal of Cryptology,* 3, 1, 1990.

[14] Hedayat, A.S., Sloane, N.J.A., and Stufken, J. *Orthogonal Arrays, Theory and Application,* Springer-Verlage, Berlin, 1999.

[15] Hu, L. On construction of optimal A^2-codes, *Northeastern Mathematical Journal,* 17, 27, 2001.

[16] Johansson, T. Lower bounds on the probability of deception in authentication with arbitration, *IEEE Transactions on Information Theory,* 40, 1573, 1994.

[17] Johansson, T. Contributions to unconditionally secure authentication, Ph.D. Thesis, Lund University, Sweden, 1994.

[18] Kirkman, T.P. On a problem in combinations, *Cambridge and Dublin Mathematical Journal,* 2, 197, 1847.

[19] König, D. Über Graphen und ihre Anwendung auf Determinantentheoric und Mengenlehre, *Mathematical Annals,* 77, 453, 1916.

[20] Lindner, C.C. and Stinson, D.R. Steiner pentagon system, *Discrete Mathematics,* 52, 67, 1984.

[21] Lindner, C.C. and Stinson, D.R. The spectrum for the conjugate invariant subgroups of perpendicular arrays, *Ars Combinatoria,* 18, 51, 1984.

[22] Massey, J.L. Cryptography – a selective survey, in *Digital Communications,* Biglieri, C. and Prati, C., Eds., Elsevier North-Holland, Amsterdam, 1986, 3.

[23] Massey, J.L. Contemporary cryptology: the science of information integrity, 1–39, in *Contemporary Cryptology: An Introduction,* Simmons, G.J., Ed., IEEE Press: New York, 1992.

[24] MacWilliams, F.J. and Sloane,N.J.A. *The Theory of Error-Correcting Codes,* North-Holland Publishing Co., Amsterdam, 1977.

[25] Mullin, R.C. et al. On the construction of perpendicular arrays, *Utilitas Mathematica,* 18, 141, 1980.

[26] Obana, S. and Kurosawa, K. A^2-code=affine resolvable +BIBD, *Lecture Notes in Computer Science 1334,* Springer-Verlag, Berlin, 1997, 118.

[27] Payne,S.E. and Thas,J.A. Finite generalized quadrangle, *Research Notes in Mathematics 110,* Pitman Publishers Inc. 1984.

[28] Pei, D. Information-theoretic bounds for authentication codes and block designs. *Journal of Cryptology,* 8, 177, 1995.

[29] Pei, D. and Wang, X. Authentication-secrecy code based on conocs over finite fields, *Science in China (Series E)*, 39, 471, 1996.

[30] Pei, D. A problem of combinatorial designs related to authentication codes, *Journal of Combinatorial Designs*, 6, 417, 1998.

[31] Pei, D. et al. Characterization of authentication codes with arbitration, *Lecture Notes in Computer Science 1587*, Springer-Verlag, Berlin, 1999, 303.

[32] Pei, D. Authentication schemes, in *Coding Theory and Cryptology*, Harald Niederreiter, Ed., Lecture Notes Series, Institute for Mathematical Science, National University of Singapore, Vol. 1, Singapore University Press and World Scientific, 2002, 283.

[33] Pei, D. and Li, Y. Optimal authentication codes with arbitration, *Acta Mathematicae Applicatae Sinica*, 25, 88, 2002.

[34] Ray-Chaudhuri, D.K. Application of the geometry of quadratics for constructing PBIB design, *Annals of Mathematical Statistics*, 33, 1175, 1962.

[35] Rosenbaun, U. A lower bound on authentication after having observed a sequence of messages, *Journal of Cryptology*, 6, 135, 1993.

[36] Safavi-Naini, R. and Tombak, L. Optimal authentication system, *Lecture Notes in Computer Science 765*, Springer-Verlag, Berlin, 1994, 12.

[37] Schöbi, P. Perfect authentication system for data sources with arbitrary statistics, Presented at *Eurocrypt'86*.

[38] Sgarro, A. Information divergence bounds for authentication codes, in *Advances in Cryptology – Eurocrypt'89, Lecture Notes in Computer Science 434*, Springer-Verlag, Berlin, 1990, 93.

[39] Sgarro, A. Information-theoretic bounds for authentication frauds, *Journal of Computer Security*, 2, 53, 1993.

[40] Shannon, C.E. Communication theory of secrecy system, *Bell System Technical Journal*, 28, 656, 1949.

[41] Simmons, G.L. Authentication theory/coding theory. *Advances in Cryptology – Crypto'84, Lecture Notes in Computer Science 196*, Springer-Verlag, Berlin, 1985, 411.

[42] Simmons, G.L. Message authentication with arbitration of transmitter/receiver disputes, *Advances in Cryptology–Eurocrypt'87, Lecture Notes in Computer Science 304*, Springer-Verlag, Berlin, 1988, 151.

[43] Simmons, G.J. A Cartesian product construction for unconditionally secure authentication codes that permit arbitration, *Journal of Cryptology*, 2, 77, 1990.

[44] Skolem, T. Some remarks on the triple system of Steiner, *Mathematica Scandinavica*, 6, 273, 1985.

[45] Smeets,B. Bound on the probability of deception in multiple authentication, *IEEE Transactions on Information Theory*, 40, 1586, 1994.

[46] Smeets, B., Vanroose, P., and Wan, Z.X. On the construction of authentication codes with secrecy and codes which stand spoofing attacks of order L at least 2, *Advances in Cryptology – EUROCRYPT'90*, Damgard, I.B, Ed., *Lecture Notes in Computer Science 473*, Springer-Verlag, Berlin, 1991, 306.

[47] Stinson, D.R. A construction for authentication/secrecy codes from certain combinatorial designs, *Journal of Cryptology*, 1, 119, 1988.

[48] Stinson, D.R. The combinatorics of authentication and secrecy codes, *Journal of Cryptology*, 2, 23, 1990.

[49] Stinson, D.R. *Combinatorial Designs: Constructions and Analysis.* Springer-Verlag, Berlin, 2004.

[50] Teirlinck, L. Nontrivial *t*-designs without repreated blocks exist for all *t*, *Discrete Mathematics*, 65, 301, 1987.

[51] Walker, M. Information-theoretic bounds for authentication codes, *Journal of Cryptology*, 2, 131, 1990.

[52] Wan, Z. *Geometry of Classical Groups over Finite Fields (Second Edition)*, Science Press: Beijing, New York, 2002.

[53] Wan, Z., Smeets, B. and Vanroose, P. On the construction of Cartesian authentication codes over symplectic spaces, *IEEE Transsctions on Information Theory*, 40, 920, 1994.

[54] Wang, Y. Authentication codes constructed by plane curves. *Advances in Cryptology – Chinacrypt'92*, Science Publication 1992, 74.

[55] Wang, Y., Safavi-Naini, R. and Pei, D. Combinatorial characterisation of *l*-optimal authentication codes with arbitration, *Journal of Combinatorial Mathamatics and Combinatorial Computing*, 37, 205, 2001.

[56] Wielandt, H. *Finite Permutation Groups,* Academic Press: New York, 1964.

[57] Wilson, R.M. Concerning the number of mutually orthogonal Latin squares, *Discrete Mathematics*, 9, 181, 1974.

[58] Witt, E. Dei fünffan tromsitiven Gruppen von Mathien, *Abhandlungen der Mathematik Hamburg*, 12, 256, 1938.

Notations

A-Codes

$p(e)$	the probability of the event that $E = e$, 8
$p(m)$	the probability of the event that $M = m$, 8
p_S	the probability distribution of the random variable S, 10
p_E	the probability distribution of the random variable E, 10
p_{S^r}	the probability distribution of the random variable S^r, 23
$P(m\|m^r)$	the probability of the event that the message m is valid of the encoding rule used by the transmitter given that the r-tuple m^r of messages has been observed, 9, 12
P_r	the expected probability of successful deception for an optimum spoofing attack of order r, 8
$H(X)$	the entropy of the random variable X, 16
$H(X,Y)$	the joint entropy of the random variables X and Y, 17
$H(X\|Y)$	the conditional entropy of X when Y is given, 17

A^2-Codes

$\mathscr{E}_T$	the set of encoding rules of the transmitter, 10
$\mathscr{E}_R$	the set of decoding rules of the receiver, 10
E_T	the random variable of encoding rules, 12
E_R	the random variable of decoding rules, 12
O_r	the spoofing attack of order r by the opponent, 11
R_r	the spoofing attack of order r by the receiver, 11
T	the spoofing attack by the transmitter, 11
$\mathscr{M}(f)$	the set of valid messages of the decoding rule f, 11
$\mathscr{M}(f,s)$	the set of valid messages of the decoding rule f, which can be used to transmit the source state s, 10
$\mathscr{M}'(e)$	the set of messages which are not valid of the encoding rule e, 11
$\mathscr{M}'_f(e)$	the set of messages which are valid of the decoding rule f and not valid of the encoding rule e, 12
$\mathscr{E}_T(f)$	the set of encoding rules which are valid of the decoding rule f, 11
$\mathscr{E}_T(f,m^r)$	the set of encoding rules which are valid of the decoding rule f, and of which the r-tuple m^r of messages is valid, 11
$\mathscr{E}_R(e)$	the set of decoding rules of which the encoding rule e is valid, 11

$E_T(f)$	the random variable of valid encoding rules of the decoding rule f, 50
$E_R(e)$	the random variable of decoding rules of which the encoding rule e is valid, 50
$\mathscr{E}_R(m^r)$	the set of decoding rules f for which $f(m^r) = \{f(m_1), f(m_2), \cdots, f(m_r)\}$ is a r-tuple of different source states, 11
$\mathscr{E}_T(m^r)$	the set of encoding rules of which the r-tuple m^r of messages is valid, 57
$\mathscr{E}_R(e, m')$	the set of decoding rules of which the given encoding rule e and the given message m' are valid, and the message m' is not valid of e, 12
$\overline{\mathscr{M}_R^r}$	the set of r-tuples m^r of messages for which the set $\mathscr{E}_R(m^r)$ is not empty, 41
$\mathscr{M}_{f,i_1,i_2,\cdots,i_r}$	the set of r-tuples $m_f^r = (m_{i_1}, m_{i_2}, \cdots, m_{i_r})$ of messages such that $m_{i_h} \in \mathscr{M}(f, s_{i_h})$ $(1 \leqslant i \leqslant r)$ for a given r-tuple $(s_{i_1}, s_{i_2}, \cdots, s_{i_r})$ of source states, 56
P_{O_r}	the expected successful probability for the optimal spoofing attack of order r by the opponent, 11
P_{R_r}	the expected successful probability for the optimal spoofing attack of order r by the receiver, 11
P_T	the expected successful probability for the optimal spoofing attack by the transmitter, 11
$P(m|f, m^r)$	the probability of the event that the message m is valid of the encoding rule used by the transmitter given the decoding rule f and the first r messages m^r sent by the transmitter, 12
$P(m'|e)$	the probability of the event that the message m' is accepted by the receiver, but it is not valid of the given encoding rule e, 12
p_{E_R}	the probability distribution of the random variable E_R, 13
p_{E_T}	the probability distribution of the random variable E_T, 13

Authentication/Secrecy Codes

s_r'	a set of unordered r distinct source states, 187
m_r'	a set of unordered r distinct messages, 187
$\mathscr{S}_r$	the set of r-subsets (unordered) of distinct source states, 187
S_r	the random variable of r-subsets of distinct source states, 187

p_{S_r}	the probability distribution of the random variable S_r, 205
$\mathscr{E}(m'_r)$	the set of encoding rules of which the r-subset m'_r is valid, 187
$\overline{\mathscr{M}_r}$	the set of those r-subsets m'_r of distinct messages such that the set $\mathscr{E}(m'_r)$ is not empty, 187
M_r	the random variable of the first r distinct messages (unordered) sent by the transmitter, 187
$U(t)$	the t-fold unordered perfect secrecy, 187
$O(t)$	the t-fold ordered perfect secrecy, 188
$S(t)$	the t-fold Stinson perfect secrecy, 189
$\mathscr{E}(s'_r, m'_r)$	the set of encoding rules which encode s'_r into m'_r, 196

Combinatorial Designs

$\mathbb{F}_q^{(n)}$	the n-dimensional vector space over $\mathbb{F}_q$, 28
$AG(n, \mathbb{F}_q)$	the n-dimensional affine space over $\mathbb{F}_q$, 116
$PG(n, \mathbb{F}_q)$	the n-dimensional projective space over $\mathbb{F}_q$, 29
$PGL_{n+1}(\mathbb{F}_q)$	the group of projective transformations of $PG(n, \mathbb{F}_q)$, 75
$PO_{n+1}(\mathbb{F}_q)$	the subgroup of $PGL_{n+1}(\mathbb{F}_q)$ which carry the curve $\mathcal{C}$ into itself, 81
$N(n, t)$	the number of t-dimensional subspaces in $PG(n, \mathbb{F}_q)$, 30
$Q_n(t, r)$	the number of r-dimensional subspaces contained in a fixed t-dimensional subspace of $PG(n, \mathbb{F}_q)$, 31
t-(v, k, λ)	a t-design, 24
t-$(v, b, k; \lambda, 0)$	a partially balanced t-design, 24
t-$(v, b, k; \lambda_1, \lambda_2, \cdots, \lambda_t, 0)$	a strong partially balanced t-design, 24
$OA(N, k, n, t)$	an orthogonal array with N rows, k columns, n levels, and strength t, 35
(v, k, λ)	a balanced incomplete block design, 117
$STS(v)$	a Steiner triple system of order v, 117
$(X, \circ)$	a pseudo-group, 119
$MOLS(n, w)$	a set of mutually orthogonal w Latin squares of order n, 128
$TD(k, n)$	a transversal design, 135
$PA(k, s)$	a perpendicular array of order k and depth s, 200
$CPA(v, s)$	a cyclic perpendicular array, 219
$\text{dist}(u, v)$	the Hamming distance between two vectors u and v, 145
$(k, N, d)_s$	a code of length k, size N, and minimal distance d over an alphabet of size s, 145

(k, N, d)	a code of length k, size N, and minimal distance d over $\mathbb{F}_q$, 145
$d^\perp$	the dual distance of a linear code (k, N, d), 146
$C^\perp$	the dual code of the linear code C, 146
$GL_{2v}(\mathbb{F}_q)$	the group of $2v \times 2v$ nonsingular matrices over $\mathbb{F}_q$, 153
$Sp_{2v}(\mathbb{F}_q)$	the symplectic group of degree $2v$ over $\mathbb{F}_q$, 153
(m, s)	a type of subspace in the $2v$-dimensional symplectic space over $\mathbb{F}_q$, 155
$P^\perp$	the dual subspace of the subspace P, 156
$\mathcal{M}(m, s; 2v)$	the set of subspaces of type (m, s) in the $2v$-dimensional symplectic space over $\mathbb{F}_q$, 161
$N(m, s; 2v)$	$|\mathcal{M}(m, s; 2v)|$, 161
$< u, v >$	the space spanned by vectors u and v, 169
$U_n(\mathbb{F}_{q^2})$	the unitary group of order n over $\mathbb{F}_{q^2}$, 179

Index